Black-and-white print of Discovery, *a painting by Robert Tabor (1962) showing (left to right) Carl E. McIlwain, James A. Van Allen, George H. Ludwig, and Ernest C. Ray studying a paper-chart record of radiation data from* Explorer I. *(Courtesy Museum of Art, University of Iowa.)*

ORIGINS OF *Magnetospheric Physics*

JAMES A. VAN ALLEN
Department of Physics and Astronomy
University of Iowa

Smithsonian Institution Press
Washington, D.C. *1983*

TO *Abigail*

© 1983 Smithsonian Institution. All rights reserved.
Printed in the United States of America.

Library of Congress Cataloging in Publication Data

Van Allen, James Alfred, 1914–
  Origins of magnetospheric physics.

  Bibliography: p.
  1. Magnetosphere—Research—History. 2. Astronautics
in geophysics—History.   I. Title.
QC809.M35V35   1983   538'.766   83-600127
ISBN 0-87474-940-9

The Preface and Table of Contents from *Scientific Uses of
Earth Satellites,* edited by James A. Van Allen, © 1956,
the University of Michigan Press, is reprinted in chapter
5 by permission of the University of Michigan Press.

Contents

Preface
6

Acknowledgments
8

I. Scientific Heritage
9

II. The U.S. Program of Rocket Flights of Scientific Equipment
15

III. Rockoon Flights from Baffin Bay to the Ross Sea
21

IV. Remarks on the Period 1946–57
31

V. Plans for Scientific Work with Artificial Satellites of the Earth
33

VI. *Sputnik I*
43

VII. Discovery of the Inner Radiation Belt of the Earth—*Explorers I* and *III*
49

VIII. The Argus Tests
73

IX. Early Confirmations of the Inner Radiation Belt and Discovery of the Outer Radiation Belt—*Explorer IV* and *Pioneers I, II, III, IV,* and *V*
85

X. Related Work with *Sputniks II* and *III* and *Luniks I, II,* and *III*
93

XI. Second Generation Investigations and Advances in Physical Interpretation
101

APPENDIXES

A. Proposal for Cosmic Ray Observations in Earth Satellites, 1955
122

B. Correspondence Regarding *Pravda* Article, 1959
128

Bibliography
131

Preface

The subject matter of this monograph is centered on the discovery of the radiation belts of the earth in early 1958 with equipment on the earth satellites *Explorers I* and *III*, on the prompt confirmations of this discovery, and on its extensions to form the basis of the now well-developed discipline of magnetospheric physics. As of 1983 magnetospheric physics is a massive and flourishing science that engages the efforts of over a thousand investigators in at least twenty different countries. The current rate of publication in this field is of the order of two original research papers per day.

The general magnetic field of the earth is essential to the existence of its magnetosphere. Indeed, the magnetosphere of the earth or of any other celestial body may be defined as that region surrounding the body within which its magnetic field, however distorted by external currents, controls the motion of electrically charged particles. It is now known that the magnetosphere of the earth encompasses a huge population of such particles—electrons, protons, and other ions—whose source and gross dynamics are traceable principally, but not wholly, to the solar wind, a magnetized, ionized gas emitted by the sun and flowing outward through the solar system.

On the one hand magnetospheric physics is closely related to laboratory plasma physics; on the other hand it is presumably applicable to natural phenomena in planetary and astrophysical systems throughout the universe. The theory of ionized gases in magnetic and electric fields provides the unifying principles of the subject over a vast range of physical scale. The present sophistication of magnetospheric physics stands in stark contrast to its primitive beginnings.

The observational work that led to the discovery of the radiation belts of the earth, later recognized as a portion of the larger magnetospheric system, was done between 1946 and 1957 with equipment on rockets fired more-or-less vertically to altitudes of the order of 100 kilometers. The scientific context of the discovery had a much longer and more diverse history.

My account is restricted to sketching the origins of magnetospheric physics and makes no pretense of describing the current state of knowledge of the subject or its planetary and astrophysical ramifications. The account has an autobiographical thread and depends on a large collection of personal journals, notebooks, correspondence, unwritten recollections, and other unpublished material for its general spirit and organization. Nonetheless, almost all of the substance is referenced to published papers that are accessible to an interested reader. Full citations of all such original sources are given. I have selected, with a few exceptions, the early 1960s as the approximate cutoff date of my account for two reasons. First, the principal elements of magnetospheric physics were moderately well recognized by then. Second, relevant investigations proliferated at such a rapid pace thereafter that I felt unable to do them justice within the time that I had available for this work.

The appended bibliography lists relevant reference works, some of relatively recent date, and original papers, selected to represent the mainstream of the early development of the subject. Among the reference works, Wilmot N. Hess's *The Radiation Belt and Magnetosphere* (1968) will be found most helpful in supplying additional substance to chapter 11 of my brief account.

As of 1981 the magnetospheric properties of the moon and the planets Mercury, Venus, Mars, Jupiter, and Saturn had been investigated *in situ* by instruments carried by Explorer, Apollo, Mariner, Pioneer, Lunik, Venera, Mars, and Voyager spacecraft. Corresponding investigations of Uranus and Neptune in 1986 and 1989, respectively, are realistic objectives of *Voyager 2*. Those of us who have participated in this work count the period that began in 1958 as one of the most rewarding in scientific history.

Acknowledgments

Most of this monograph was written between January and August 1981 while I was a Regents' Fellow of the Smithsonian Institution, on leave from my regular teaching and research duties at the University of Iowa. The fellowship was kindly arranged by Dr. Noel Hinners, then director of the National Air and Space Museum of the Smithsonian Institution. I am greatly indebted to him, to Drs. Paul A. Hanle and David H. DeVorkin, and to others of the NASM staff for many courtesies and much help. At the University of Iowa Evelyn D. Robison provided highly competent editorial and bibliographic assistance through many versions of the manuscript, and John Birkbeck provided skilled assistance with the illustrations.

It is impossible for me to adequately acknowledge my indebtedness to the uncounted colleagues and students who participated in the experimental work on which this monograph is based. I may only hope that their roles are properly represented in the text.

Finally and most importantly, I acknowledge the unfailing support and encouragement that I have received over many years from my wife, Abigail, to whom this monograph is dedicated, and from our children, Cynthia, Margot, Sarah, Thomas, and Peter.

I.
Scientific Heritage

The scientific heritage of magnetospheric physics lies principally in studies of geomagnetism, aurorae, and the geophysical aspects of cosmic radiation and solar corpuscular streams. The external magnetic field of the earth plays a central role in the phenomena of all of these subjects. It was shown by Gilbert [1600] that the magnetic field on the surface of the earth is similar to that on the surface of a uniformly magnetized sphere (terrella) of magnetite, or lodestone. Alternatively, such a magnetic field can be attributed [Maxwell 1891] to a point magnetic dipole at the center of a nonmagnetic sphere or to a small current-carrying loop of wire there. The magnetic field external to the sphere (fig. 1) is identical to all three cases and does not distinguish among them. Geophysical evidence, however, shows conclusively that the general magnetic field of the earth must be attributed to a system of electrical currents in its deep interior [Chapman and Bartels 1940; Rikitake 1966].

By the use of networks of fixed and portable magnetometers, refined studies of the geomagnetic field as a function of time and as a function of latitude and longitude over the surface of the earth reveal an immense richness of detail. The simplest approximate representation of the general field is that of a point magnetic dipole of moment 8.0×10^{25} gauss cm^3, located at the geometrical center of the earth with its axis tilted to the rotational axis of the earth by $11°.4$ such that it pierces the earth's surface at latitude $78°.6$ N, longitude $69°.8$ W in the northern hemisphere, and at latitude $78°.6$ S, longitude $110°.2$ E in the southern hemisphere. In the conventional nomenclature of magnetic poles, the north pole of the dipole lies in the southern hemisphere of the earth and vice versa. In a next higher order representation, the point dipole retains the same orientation but is displaced from the geometrical center of the earth by 450 km toward latitude $17°.2$ N, longitude $148°.8$ E. The latter (eccentric dipole) representation is reasonably adequate for most magnetospheric purposes at low latitudes ($\pm 45°$) and at radial distances less than about 5 R_E (1 R_E = 6378 km, the equatorial radius of the body of the earth), though the addi-

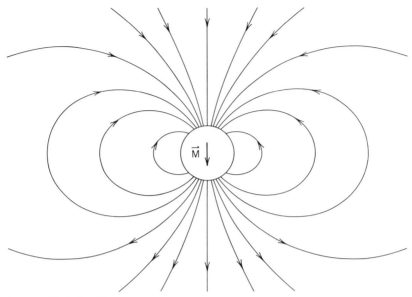

Courtesy of J. A. Van Allen.

Fig. 1. Schematic diagram of the idealized dipolar magnetic field external to a sphere, at the center of which is a loop of current-carrying wire having a magnetic moment $\vec{M}$.

tion of higher order terms is required to reproduce the effects of local anomalies. At large radii a much more elaborate description of the external magnetic field of the earth is required.

Superimposed on the average field at the surface of the earth, relatively small (typically $\lesssim 1\%$ in magnitude and $\lesssim 1°$ in direction) but highly significant temporal variations in the magnetic vector occur. From the point of view of magnetospheric physics, the most important temporal variations (magnetic "storms") occur sporadically but also exhibit a quasi-persistent periodic variation with the synodic period of rotation of the sum (~27 days). The onsets of magnetic storms are identified with flares in localized, disturbed regions on the sun and are a delayed effect, the delay typically being two days. During the first half of the twentieth century [Chapman and Bartels 1940], this evidence was taken to imply the sporadic emission of solar corpuscular streams (streams of ions and electrons, or neutral but ionized gas) traveling radially outward through the vacuum of interplanetary space at a velocity of as much as 900 kilometers per second, as calculated from the time delay. The arrival of such a stream at the earth was visualized as being the cause of a magnetic storm by first compressing the terrestrial field (initial phase) and by then generating a westward-flowing, equatorial ring current encircling the earth (main phase).

The twenty-seven-day recurrence of magnetic disturbances was taken to signify the quasi-persistence of such a corpuscular stream, having a localized spread in solar longitude and latitude and an intensity that decayed with a lifetime of the order of several months. Often, two or more such hypothetical streams were present simultaneously, causing a superposition of geomagnetic effects. By virtue of the rotation of the sun and the assumed radial motion of the stream, an individual stream would have the instantaneous shape of an archimedean spiral, as viewed in an inertial coordinate system centered on the sun (fig. 2). The quantitative aspects of this theory were developed most notably by Chapman and Ferraro [1931, 1932] and, from a different point of view, by Alfvén [1950, 1955]. Much of this early work continues to be a valid and basic part of contemporary concepts.

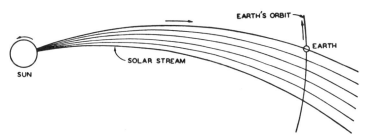

Courtesy of Oxford University Press.

Fig. 2. Sketch of the geometrical form of a stream of corpuscles (particles) emitted nearly radially from the rotating sun at such a speed as to reach the earth after a travel time of thirty-six hours. The view is from the north celestial pole [Chapman and Bartels 1940]. Later evidence shows that streams of particles of lesser speed, and hence of more sharply curved form, are much more common.

Another important body of relevant knowledge came from systematic study of the polar aurora, the common and often spectacular display of luminous emissions in the upper atmosphere at high latitudes, both north and south. The occurrence of overhead auroral displays is observed to be most probable, not at the magnetic poles, but within two halo-shaped strips (the auroral zones or auroral ovals) encircling the earth, one in the northern hemisphere and one in the southern hemisphere. These two strips are approximate mirror images of each other with respect to the geomagnetic equator; they have a latitudinal width of about 10° and are centered on the average at geomagnetic latitudes 67° north and south (fig. 3). Auroral emissions occur most prominently at altitudes of 60 to 200 km. The spec-

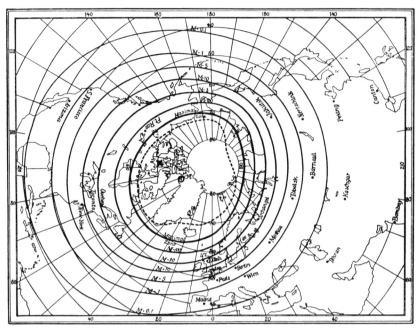

Courtesy of Oxford University Press.

Fig. 3. A diagram after H. Fritz (1881) showing lines of equal frequency of occurrence of visible aurorae. The unit of frequency M is number of nights per year. The curve labeled "Maximal Zone" defines the auroral oval whose approximate center is the geomagnetic pole near Thule, Greenland. Along the dashed curve, aurorae "occur equally often in the northern and the southern" sky [Chapman and Bartels 1940].

trum of auroral light is principally that of excited neutral and partially ionized atoms of the major atmospheric constituents, nitrogen and oxygen. In early work the exciting agent was usually assumed to be downward streaming electrons having such energies as would permit them to penetrate the atmosphere to an altitude of 60 km, namely about 10 keV. Birkeland [1908, 1913] conducted a major observational program on the aurora polaris and then a series of laboratory investigations with electron beams in the field of a small magnetized sphere (terrella) placed in an imperfect vacuum. In these laboratory studies he demonstrated the scaled occurrence of auroral ovals, resembling those of the earth. Later investigations of this nature are exemplified by the work of Malmfors [1945] and Block [1955]. Birkeland's experiments stimulated Størmer [1955] to embark on a long career devoted to the theory of the motion of electrically charged particles in the field of a magnetic dipole. This theory never succeeded in explaining the observed auroral ovals as being caused by the arrival of either electrons or ions directly from the sun, but it provided a great stimulus to the subject. Perhaps even more important, it laid the basis for understanding two related but different bodies of phenomena: (*a*) the role of the earth's dipolar field as a huge magnetic spectrometer for cosmic rays and for energetic particles from the sun and (*b*) the durable trapping of charged particles within the geomagnetic field.

Størmer's theory of the allowed cone for the arrival of energetic charged particles from infinity at a given latitude in a given direction (later refined by Lemaitre, Vallarta, and many others) is basic to deriving the energy spectrum of the primary cosmic radiation from the dependence of intensity on latitude, to determining the algebraic sign of the electrical charge on the arriving particles, and hence to distinguishing electrons from ions.

Størmer's work has been the foundation of much of my own research efforts over a period of thirty-seven years. In its simplest form his theory shows that, for an isolated, electrically charged particle of specified magnetic rigidity $R \equiv pc/Ze$ (where p is its momentum; Ze, its electrical charge; and c, the speed of light), there are, subject to certain restrictions on the magnitude of R, two dynamical regions within a dipolar magnetic field—one of unbounded motion and one of bounded motion. The first is the one of interest for the arrival of particles from infinity; the second, the one that is basic to the physics of geomagnetically trapped particles (fig. 4). In the

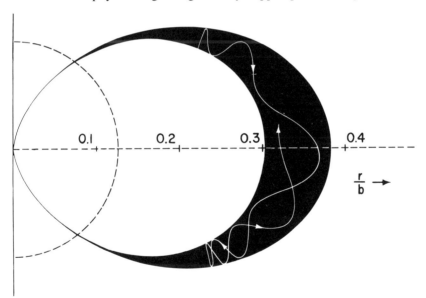

Courtesy of Archives des Sciences, Muséum d'Histoire Naturelle, Geneva.

Fig. 4. A diagram, after Størmer [1907], illustrating the meridian projection of the spatial trajectory of an energetic, electrically charged particle in the field of a magnetic dipole and the boundaries of the theoretically rigorous trapping region. The quantity r/b is the dimensionless ratio of the radial distance to a parameter of Størmer's theory of this motion. The earth is represented by the dashed semicircle.

simple case of the motion of isolated, noninteracting particles in a vacuum, there is no connection between the two regions—i.e., a particle arriving near the earth from infinity must either pass by the earth or strike the atmosphere but can never become trapped whereas one injected into the trapping region can never escape therefrom, though in the real case it might collide with the atmosphere and be lost. As will be shown later, much of the physics of the magnetosphere results from departures from the idealized Størmerian case. Nonetheless, the simple theory provides a point of departure for all more elaborate treatments of the subject.

Beginning in the 1930s, Jacob Clay, Arthur H. Compton, Robert A. Millikan, Ira S. Bowen, H. Victor Neher, William H. Pickering, Erich Regener, Georg Pfotzer, Hugh Carmichael, and others undertook latitude surveys of cosmic-ray intensity using ionization chambers and Geiger tube detectors at ground level and carried by balloons to altitudes up to about 30 km [Bowen, Millikan, and Neher 1938]. Their measurements (fig. 5) were of basic importance, but they left the nagging question of how to extrapolate reliably the highest altitude measurements, still beneath some 10 grams per square centimeter of atmosphere, to free space, i.e., above the appreciable atmosphere. For many years the desire to answer this question was the central motivation for much of my scientific work, leading fortuitously to the first direct observations of the primary auroral radiation and to the discovery of the radiation belts of the earth.

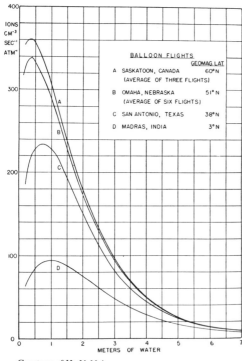

Fig. 5. The dependence of cosmic-ray ionization on depth in the earth's atmosphere, measured in equivalent meters of water, at four different geomagnetic latitudes [Bowen, Millikan, and Neher 1938].

Courtesy of H. V. Neher.

II.

The U. S. Program of Rocket Flights of Scientific Equipment

The preceding pages give a brief sketch of certain elements of the solar-geophysical knowledge that was available before the start of high-altitude observations with rocket-borne equipment in 1946.

As early as 1943 Erich Regener and Ernst Stuhlinger in Germany planned investigations of the cosmic-ray intensity above the atmosphere and of the solar ultraviolet spectrum with equipment carried on test flights of V-2 military rockets. But such flights were never conducted because of wartime operations, including Allied air raids on the German rocket base at Peenemünde that destroyed some of the instruments already built for the purpose.

The post–World War II period was characterized by intensive efforts within the United States and the Soviet Union to develop high-performance rockets for military purposes. At a less well-known level it was also characterized by the aspirations of scientists to use such rockets as vehicles for carrying scientific equipment to high altitudes. On September 26, 1945, the Jet Propulsion Laboratory (JPL) successfully flew a small sounding rocket, called a WAC Corporal, from the newly established White Sands Proving Ground (WSPG) in New Mexico. This rocket, a reduced-scale version of the JPL military rocket Corporal, reached an altitude of 70 km with a potential payload of 5 kg. In some sense, this flight advanced the much earlier plans of Robert Goddard to conduct scientific work with rockets, flown to high altitudes for sounding, or investigating, the upper atmosphere; but the WAC Corporal was not applied in any significant manner to scientific work.

The first major development in the history of the rocket flight of scientific equipment came a few months later. In late 1945 the U.S. Army Ordnance Department transferred a group of German rocket engineers and a large stock of V-2 rocket components from Peenemünde to the United States. It planned to assemble and fire a number of V-2s for the purpose of technical assessment and experience as part of the then embryonic effort of the United States to develop ballistic missiles. In response to the expressed interest of Ernst H. Krause of the Naval Research Laboratory (NRL), Col. Holger N. Toftoy and Lt. Col. James G. Bain invited scientists from universities and government laboratories to formulate a program for the utilization of the payload capacity ($\sim$1,000 kg) of the V-2 test flights for the conduct of scientific investigations [Newell 1953].

At that time I was seeking to return to civilian employment at the Applied Physics Laboratory (APL) at Johns Hopkins University after service as an ordnance and gunnery officer in the U.S. Navy since November 1942. In 1940 and 1941 I had worked at the Department of Terrestrial Magnetism (DTM) of the Carnegie Institution of Washington in helping develop rugged vacuum tubes and photoelectric and radio proximity fuzes for gun-fired projectiles. The director of DTM, Merle A. Tuve, had established APL in early 1942 and then served as its wartime director. Along with other colleagues in the proximity fuze group at DTM, I was transferred to APL at the time of its creation. At Tuve's request I was commissioned directly from my post at APL into the U.S. Naval Reserve as a lieutenant (junior grade) in November 1942 as one of three individuals to take the first issue of secret radio-proximity fuzed 5"/38 projectiles to the South Pacific Fleet. My functions were to introduce this new type of antiaircraft ammunition into service in the fleet, to instruct gunnery officers in its use, to observe and report its effectiveness in combat, and later to set up re-batterying facilities in Australia, New Caledonia, Espíritu Santo, Tulagi, Eniwetok, and Manus. After my return to the United States and the end of the war in the Pacific, I had a number of conversations with Tuve on resuming peacetime research. My 1939 Ph.D. from the University of Iowa had been in experimental nuclear physics, and I had continued in this field at DTM as a research fellow of the Carnegie Institution in 1939–40. During this period I had acquired an interest in cosmic-ray research from Scott Forbush and others at DTM. Tuve encouraged me to pursue this interest and invited me to return to APL to conduct research in this field.

Henry H. Porter of APL told me about the Army Ordnance Department's plans for the V-2 firings, and on January 16, 1946, I joined with many other interested scientists at a meeting at the Naval Research Laboratory for a briefing on the possibilities. One tangible result of this meeting was the formation of an unofficial group of scientists who had a realistic expectation of preparing equipment for flight. I was fortunate enough to be a member of this group, a circumstance that was pivotal to my subsequent career. My gunnery experience and my earlier intensive experience in developing vacuum tubes and associated circuits that survived linear accelerations of up to 20,000 g while being fired from guns led me to think that building electronics and scientific instruments for rocket-borne payloads would be easy—an expectation that later proved to be only partially true.

With Tuve's support I organized a high-altitude research group of kindred spirits: Howard E. Tatel, John J. Hopfield, Robert Peterson, Lawrence W. Fraser, Russell S. Ostrander, Clyde T. Holliday, and, later, Jeffrey F. R. Floyd, S. Fred Singer, Gene M. Melton, Albert V. Gangnes, James F. Jenkins, Jr., Harold E. Clearman, and others. We very quickly developed plans for a comprehensive program of measurements of primary cosmic rays, the ultraviolet spectrum of the sun, and the geomagnetic field in the ionosphere. Ernest H. Vestine of DTM and Allen Maxwell of the Naval Ordnance Laboratory were important collaborators in the high-altitude magnetometer program for the direct measurement of the magnetic effects of ionospheric currents. Later, the solar ultraviolet work led Hopfield and me into a determination of the altitude distribution of ozone in the upper atmosphere. Also, Holliday began the development of cameras with recoverable film casettes for photographing the earth from high altitudes, in both the visible and photographic infrared. In addition to general supervision of this work, my own special interest was investigating the nature and absolute intensity of the primary cosmic radiation before it encountered the earth's atmosphere.

The nationwide group of scientists and engineers from a diversity of universities and federal and industrial laboratories, as mentioned above, had no official status or authority but was entrusted by the Army Ordnance Department with developing a scientific program, allocating available payload capacity on V-2 test flights, and advising it on such other matters as would assure successful execution of the program. This self-constituted group called itself the "V-2 Upper Atmosphere Panel"; its initial members were [Newell 1980]:

- E. H. Krause, Naval Research Laboratory (chairman)
- C. F. Green, General Electric Company
- K. H. Kingdon, General Electric Company
- G. K. Megerian, General Electric Company (secretary)
- M. H. Nichols, Princeton University
- F. L. Whipple, Harvard College Observatory
- W. B. Dow, University of Michigan
- M. J. E. Golay, Signal Corps Engineering Laboratory
- J. A. Van Allen, Applied Physics Laboratory/Johns Hopkins University
- M. D. O'Day, Air Force Cambridge Research Laboratory
- N. Smith, National Bureau of Standards

The first formal meeting of the panel occurred on February 27, 1946. Work on preparing scientific instruments was already underway, and a provisional schedule of launchings was agreed upon. Research and development proceeded rapidly, as was evidenced by the first three V-2 flights in the United States being conducted at the newly established White Sands Proving Ground in New Mexico on April 16 (vehicle failure), May 10, and May 29, 1946. These flights carried cosmic ray equipment prepared by the APL group, but no scientific results were obtained because of a variety of technical failures in the equipment, including the destruction, upon impact, of a magnetic wire recorder. The assembly of the vehicles was performed by a mixed group of engineers and technicians from the General Electric Company, supervised by L. D. White, and from the German Peenemünde group, supervised by Wernher von Braun. All of the work, including the fueling and launching of the rockets and the radio tracking and telemetry receptions, was performed under rather primitive conditions. The Naval Research Laboratory built the payload shells for all early flights and provided telemetry transmitters, tracking beacons, and command receivers. George Gardiner's group at the New Mexico College of Agriculture and Mechanic Arts in nearby Las Cruces performed tracking, telemetry reception, and many other valuable services.

Despite a comprehensive variety of mistakes of omission and commission in these very exciting and demanding circumstances, we pressed on with our programs, learning rapidly by diagnosis of our failures.

In 1947 Krause left the NRL for employment in industry, and I was elected chairman of the "V-2 Upper Atmosphere Research Panel," as it was then called. The panel, whose membership varied with the passage of time, met every one to three months, rotating its meeting site among offices in Washington, D.C., and the participating laboratories. The meetings were a mixture of shared experiences, plans, and results and a continuous updating of schedules and assignments of payload space. For over a decade the panel (later called the "Upper Atmosphere Rocket Research Panel" and, after April 1957, the "Rocket and Satellite Research Panel") was the focal point of the entire effort of the United States in what was later called space research [Newell 1980]. The well-recorded minutes of its some seventy-six meetings provide the working history of the subject. I continued as chairman of the panel until 1958, at which time the National Aeronautics and Space Administration and the Space Science Board of the National Academy of Sciences were created and the panel ceased to have a significant role in national planning and coordination.

Because of the limited supply of V-2s and the great expense of reproducing and firing them, it was soon realized that a much simpler, less expensive rocket would be necessary for the continuation and expansion of scientific studies at high altitudes. The APL initiated such a development by the Aerojet Engineering Corporation and the Douglas Aircraft Company in 1947. The development was supported by the Navy Bureau of Ordnance and the Office of Research and Inventions (later the Office of Naval Research). In consultation with Rolf Sebersky of Aerojet, I specified the desirable characteristics (90 kg of payload within an ogival shaped envelope 38 cm in diameter and 160 cm in length to 100 km altitude), later modified somewhat, and supervised the development of the vehicle, the design of a suitable launching tower at WSPG, and the development of telemetry, antennas, and other necessary basic support equipment for scientific instrumentation. I also oversaw the development of a variety of investigations in cosmic rays, high-altitude photography of the earth, the solar ultraviolet spectrum, the distribution of ozone in the upper atmosphere, and the magnetic field of electrical currents in the ionosphere.

The first fully powered flight of the two-stage Aerobee, as we named this new rocket, was made from WSPG on November 24, 1947. The summit altitude was only 58 km, however, because of command cut-off of the thrust of the second stage as it drifted out of our range safety grid. The second flight on March 5, 1948, was both technically and scientifically successful in all respects, reaching a summit altitude of 113 km [Van Allen, Fraser, and Floyd 1948]. Subsequent Aerobee flights were interleaved with V-2 flights and were made at the rate of about six per year. (As of 1982 over one thousand Aerobee rockets, in the original version and several upgraded versions [Van Allen, Townsend, and Pressly 1959; Townsend et al. 1959], have been flown for a wide variety of scientific investigations in upper atmospheric physics and in astronomy.)

Because of their simplicity and low cost, Aerobees became the first U.S. high-altitude rockets to be launched for scientific purposes at a geographic location other than WSPG. In 1949 the APL group conducted two successful Aerobee flights from the U.S.S. *Norton Sound* off the coast of Peru near the geomagnetic equator, and in late 1950 the same group conducted two successful Aerobee flights in the Gulf of Alaska, also from the U.S.S. *Norton Sound*. These expeditions pioneered the extension of scientific rocketry to a wide range of geographic locations. We obtained the first observations of cosmic-ray intensity above the atmosphere at two points that were widely spaced in latitude and bracketed the White Sands Proving Ground. This three-point survey permitted the construction of an approximate energy spectrum of the primary cosmic radiation [Van Allen and Singer 1952]. Also, we were successful in obtaining for the first time the magnetic signature of the equatorial electrojet in the lower E-layer of the ionosphere [Singer, Maple, and Bowen 1951]. (An example of early results from an Aerobee flight is shown in figure 6. Samples of results from our other V-2 and Aerobee flights are contained in Van Allen and Tatel 1948; Van Allen and Gangnes 1950a,b; Holliday 1950; and Hopfield and Clearman 1948.)

Leo Goldberg well characterized high-altitude research during the 1940s and 1950s when he remarked, many years later, that one of the principal products, if not one of the purposes, of space research was character building. His comment applied to both the participants and their often bereft families.

The immense opportunity for finally being able to make scientific observations through and above the atmosphere of the earth drove us to heroic measures and into a new style of research, very different than the laboratory type in which many of us had been trained. In my own case familiarity with nuclear physics techniques, exposure to cosmic-ray and geophysical problems at DTM, and World War II experience in ordnance

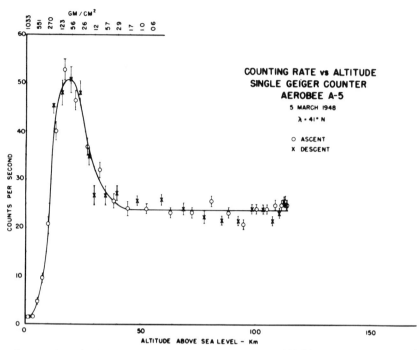

Courtesy of the American Physical Society, 1 Research Road, Box 1000, Ridge, NY 11961.

Fig. 6. Cosmic-ray intensity as a function of altitude over the White Sands Proving Ground in New Mexico. Note the essentially constant intensity above 50 km (the high-altitude plateau) [Gangnes, Jenkins, and Van Allen 1949].

and gunnery combined to impel me into this new field of research.

A much fuller account of the early U.S. program of rocket flights than I have attempted herein can be found by referring to the excellent books written or edited by Homer E. Newell, Jr. [1953, 1959, 1980].

III.

Rockoon Flights from Baffin Bay to the Ross Sea

In December 1950 I left the APL and became professor of physics and head of the Department of Physics (since 1959, the Department of Physics and Astronomy) of the University of Iowa, my Ph.D. alma mater. I immediately initiated a program of balloon flights for cosmic-ray composition studies with the help of a grant from the Research Corporation. Soon thereafter, I prepared a proposal to the U.S. Office of Naval Research (ONR) for a comprehensive study of the latitude dependence of cosmic-ray intensity and for other purposes using small rockets launched from balloons at altitudes of about 17 km. This "rockoon" technique was envisioned as an inexpensive method for sending scientific equipment to altitudes of the order of 100 km. Launching from shipboard was especially desirable for several reasons: (*a*) a ship could steam downwind to produce a zero relative wind to ease the launching, (*b*) a ship at sea could avoid populated areas and minimize the possibility of damage by the falling rockets, and (*c*) a ship could readily cover a large range of latitude, carrying the launching and telemetry support equipment with it. The general scheme for launching rockets from high-latitude balloons was suggested to me by Lt. Lee Lewis, USN, and further developed in wardroom conversations with him, Lt. Cdr. George Halvorson, USN, and S. F. Singer during the Aerobee firing cruise of the U.S.S. *Norton Sound* in March 1949. The basic feature of the scheme was the avoidance of aerodynamic drag on the rocket in the dense lower atmosphere and, hence, the achievement of much higher altitudes than those that were possible from a ground-based launcher.

(F. Winter [private communication 1981] has called my attention to a one-page illustration with caption in the popular magazine *Modern Mechanix and Inventions* of July 1934 in which the flight of a manned balloon-rocket is depicted. The only information on the scheme is contained in the caption, which reads: "A balloon-rocket conceived by a Wyoming inventor is expected to reach 43 miles into the stratosphere. Carried 11 miles by the balloon the operator cuts loose, ignites two opposed rockets and soars 33 miles higher. One mile is lost cutting away from the balloon. When the rocket power is expended, an air vent is opened, filling a parachute which floats the tube to earth. Inserts show release of rocket and details of the operating mechanism of the cylinder.")

At Iowa I developed the practicalities of the balloon-launched rocket (rockoon) technique with the help of Melvin Gottlieb and my students Joseph Kasper and Ernest Ray. The ONR approved my proposal, funded the work, placed it under the scientific cognizance of Urner Liddel, and provided essential operational support in many different ways. Following preliminary flight tests of various elements of the system at WSPG in June and July 1952, Lt. Malcolm S. Jones, USN, and Leslie H. Meredith, Lee F. Blodgett, and I of the University of Iowa undertook our first rockoon expedition on the U.S. Coast Guard Cutter *Eastwind* (AGB 279), an icebreaker, in the Arctic during August and September 1952 [Van Allen and Gottlieb 1954; Van Allen, 1959a] (figs. 7, 8). Five of the seven attempted

Courtesy of the U.S. Coast Guard.

Fig. 7. Launching a rockoon from the helicopter deck of the USCGC Eastwind *off the coast of Greenland in 1952.*

Fig. 8. A rockoon in flight, immediately after release. The instrumented Deacon rocket is suspended by a long load line from the balloon. The rectangular object dangling from the tail of the rocket is the firing box containing a battery, a barometric switch, and a mechanical timer (1952).

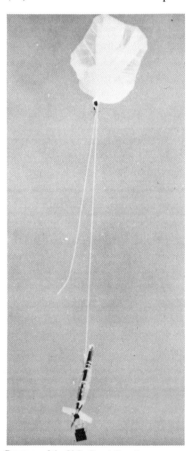

Courtesy of the U.S. Coast Guard.

flights were ballistically successful. Two of these carrying single Geiger tubes yielded reliable values of the absolute intensity of cosmic radiation up to an altitude of 90 km at a geomagnetic latitude of 88° N, and two others carried ionization chambers and gave good values of total cosmic-ray ionization up to 64 km at about the same latitude. The Geiger tube data established the essential absence of primary cosmic-ray protons in the energy interval 590 to 18 MeV [Van Allen 1953].

Between July and September 1953 the Iowa group conducted sixteen more rockoon flights of single Geiger tubes and ionization chambers (fig. 9). Seven of these flights were successful in all respects. Four carried ionization chambers, and three carried single Geiger tubes. The ionization chamber data at geomagnetic latitudes 56°, 76°, and 86° N showed that primary cosmic-ray nuclei of $Z \geq 6$ were absent for magnetic rigidities less than 1.5×10^9 volts; the result was most significant for the carbon, nitrogen, oxygen group [Ellis, Gottlieb, and Van Allen 1954], whose intensities had been well measured at greater rigidities (energies) by other investigators using balloon-borne detectors.

Courtesy of the University of Iowa.

Fig. 9. Leslie H. Meredith (left) and Robert A. Ellis, Jr. (right), with an array of flight equipment prior to a shipboard rockoon expedition to the Arctic in 1953.

Data from the Geiger tube flights confirmed and extended the 1952 results on the absence of low-energy protons in the primary cosmic-ray spectrum [Meredith, Van Allen, and Gottlieb 1955]. More important, two flights at geomagnetic latitudes 75° N and 64° N discovered a new effect above an altitude of 50 km, namely a high intensity of easily absorbable radiation superimposed on the high-altitude plateau of constant cosmic-ray intensity (fig. 10). This soft radiation was present only within the auroral zone, being absent at both higher and lower latitudes. For this reason we called it "auroral soft radiation." The observations constituted the first *in situ* detection of the primary auroral radiation [Meredith, Gott-

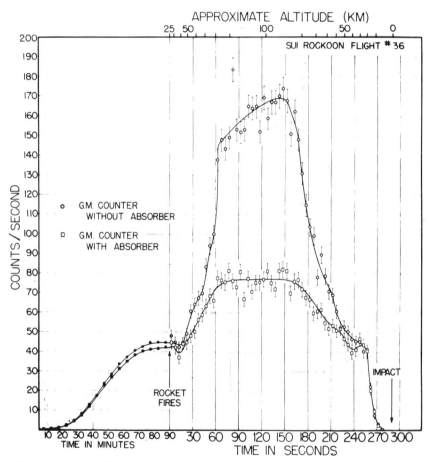

Reprinted from *Scientific Uses of Earth Satellites*, ed. James A. Van Allen, © 1956, by permission of the publisher, the University of Michigan Press.

Fig. 10. *An example of the detection of the auroral soft radiation with a rockoon flight off the southwestern coast of Greenland. Note that the time scale is in minutes during the ascent of the balloon and in seconds during the flight of the rocket* [Van Allen 1957].

lieb, and Van Allen 1955; cf. Davis, Berg, and Meredith 1960]. We showed that the particles in question were most likely electrons but of such low energy that they could not have arrived at the earth directly from a distant source (e.g., the sun). Also, we offered the tentative interpretation that the particles were part of the high-energy (~1 MeV) tail of the auroral spectrum and were directly penetrating the wall of the nose cone and the wall of the Geiger tube.

We immediately set to work preparing apparatus for further study of the latitude distribution and the nature of the auroral soft radiation. The system of detectors comprised pairs of Geiger tubes—one with the thinnest practical wall and the other with a shield of lead or aluminum—and very thinly shielded NaI (Tl) scintillation crystals equipped with current-carrying coils for magnetic deflection of electrons and mounted on photomultiplier tubes. The follow-on Arctic expeditions were made by the Iowa group in the summers of 1954 and 1955. The combination of results from these flights [Van Allen 1957] showed conclusively that the auroral primary radiation consisted of electrons having energies of the order of tens of kiloelectron volts and that the responses of the Geiger tubes were due to bremsstrahlung [Frank 1962] (x rays) produced on the nose cone by electron bombardment and *not* by directly penetrating electrons (as had been suggested earlier). The electrons were recorded directly by the scintillation crystals. Estimated absolute intensities of electrons in the primary auroral beam were in the range

10^6 to 10^8 electrons $(cm^2 \ sec)^{-1}$,

for electrons in the energy range 10 to 100 keV. The corresponding energy flux was

0.01 to 1 erg $(cm^2 \ sec)^{-1}$,

consistent with estimates based on the optical brightness of visible aurorae. An improved latitude dependence of the effect was also obtained (fig. 11). This work by Leslie H. Meredith, Melvin B. Gottlieb, Frank McDonald, Carl E. McIlwain, George H. Ludwig, Joseph C. Kasper, Ernest C. Ray, Robert A. Ellis, Jr., Jason Ellis, and me constituted the first *in situ* detection and identification of the primary auroral radiation and the first direct determination of its intensity. Our new findings were received warmly by Sydney Chapman, who was a visiting professor at the University of Iowa from October 1954 to February 1955. He lectured on "Physics and Chemistry of the Upper Atmosphere" [1955] and profoundly influenced the course of our work.

As part of our 1955 expedition, McDonald, McIlwain, Kasper, and Ludwig flew a small, two-stage rocket from a high-altitude balloon in the auroral zone with the intention of reaching an altitude of several hundred kilometers. Both stages of the rocket fired properly. The radiation detectors operated properly throughout the burning of the first stage, but the radio transmitter was modulated violently from 2.5 to 3.6 seconds after the firing of the second stage and then ceased to transmit [McIlwain 1956]. Later, we realized that we had grossly underestimated the aerodynamic heating of the aluminum nose cone of the second stage at the expected velocity of 2.4 kilometers per second. The remains of the second stage may very well have set a new altitude record in the Arctic of over 300 kilometers. Retrospectively, it appears likely that this inexpensive technique, given a heat-resistant nose cone, would have resulted in discovery of the geomagnetically trapped radiation.

The culmination of our rockoon program occurred in 1957 as part of the International Geophysical Year, with our work being supported by the National Science Foundation, by the Office of Naval Research, and operationally by the U. S. Navy.

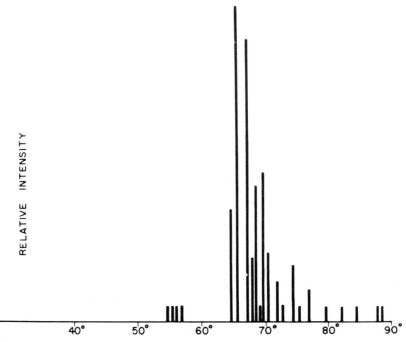

Reprinted from *Scientific Uses of Earth Satellites*, ed. James A. Van Allen, © 1956, by permission of the publisher, the University of Michigan Press.

Fig. 11. The geomagnetic latitude dependence of the maximum counting rate of lightly shielded Geiger-Mueller tubes flown in the Arctic to high altitudes by balloon-launched rockets on twenty-two occasions during the summers of 1953, 1954, and 1955. The great intensity peak centered at about 68° is attributed to primary auroral electrons having energies of $E_e \sim 10$ kiloelectron volts (the auroral soft radiation) [Van Allen 1957].

In late July 1957 Lawrence J. Cahill, Jr., Gary L. Strine, Donald E. Simanek, and I departed from Norfolk, Virginia, on the U.S.S. *Plymouth Rock* (LSD-29) for a long-planned expedition to the Arctic for the purpose of conducting a further series of rockoon flights in the northern auroral zone. Between August 5 and 14 we conducted eighteen rockoon flights in Baffin Bay, Davis Strait, and the Labrador Sea at geographic latitudes ranging from 55° to 75° N. Excellent data on the cosmic-ray intensity and the auroral soft radiation were obtained on seven successful rocket flights up to altitudes of 132 km. Cosmic-ray intensity data were obtained up to balloon altitudes of 24 km on five other flights for which the rocket failed to fire or the transmitter failed soon after rocket ignition. A nighttime rocket flight at 47° W, 56° N on August 14, 1957, was noteworthy. The rocket flew beneath and probably into a brilliant auroral display and reported one of the two highest intensities of auroral soft radiation ever observed. In fact, this nighttime flight provided the first directly observed association of visible aurora with rocket-counter results on the soft radiation. All other successful flights had been conducted during daylight conditions, with no direct knowledge of the existence of (visible) aurorae. In parallel with our rocket flights in the auroral zone, x rays in association with auroral displays were observed by other workers with balloon-borne detectors [Winckler and Peterson 1957; Winckler et al. 1958; Anderson 1958].

Five rockoon flights of Cahill's small proton precession magnetometer were also conducted as part of our July-August 1957 expedition. One of these flights at 56° W, 64° N on August 6 yielded the first measurements of the scalar geomagnetic field at high altitudes in the Arctic and the first direct confirmation and measurement of the auroral electrojet current in the altitude range 100 to 120 km [Cahill 1959b], a phenomenon previously inferred from ground-based magnetometer records.

Each of the above Hawk Rockoon flights utilized a modified Loki II solid propellant rocket (Cooper Development Company) and a 26,000-cubic-foot plastic balloon (Raven Industries, Inc.). The rocket was suspended from a long nylon load line and fired by a barometric switch at an approximate altitude of 23 km. The scientific payload had a weight of about 4 kg and was housed within a pointed cylinder of diameter either 7.6 or 8.9 cm and length 97 cm.

After a brief return to our home base in Iowa City, Cahill, Strine, and I delivered a large quantity of test equipment, telemetry receivers, scientific payloads, rockets, balloons, helium bottles, etc., to a large U.S. Navy icebreaker, the U.S.S. *Glacier* (AGB-4), in Boston harbor, in preparation for the equatorial-Antarctic phase of our IGY rockoon program.

We set up our laboratory in the after hold immediately above the screws and secured all of our equipment in order to cope with the ship's motion. The *Glacier*, having a round bottom and virtually no keel, had a reputation for rolling 15° to 20° in even a light sea. Later, we found this reputation to be well deserved, and while enroute from Antarctica to New Zealand, we experienced rolls of over 45° from the horizontal. On September 23, 1957, Cahill and I sailed with the ship from Boston, southbound via the Cape Cod Canal.

The scientific objectives of this expedition were similar in nature to those of the preceding Arctic expedition, namely (*a*) to measure the cosmic-ray intensity through and especially above the atmosphere as a function of latitude, (*b*) to determine the altitude and latitude distribution of the equatorial electrojet current system (previously identified by our Aerobee flights off the coast of Peru in 1949), (*c*) to search for soft radiation in the southern auroral zone, and (*d*) to search for ionospheric currents in the southern auroral zone. No high-altitude measurements had previously been made in the Antarctic. Points of special interest were to check the presumed north-south symmetry of the latitude dependence of the cosmic-ray intensity above the atmosphere and the incidence of soft radiation in the southern auroral zone. The rockets, balloons, and scientific payloads were essentially the same as those used in the July-August Arctic expedition.

Between September 26 and November 8, 1957, we conducted a total of thirty-six rockoon flights in the latitude range 31° N to 72° S. Good cosmic-ray data were obtained during both balloon and rocket phases of four flights (the most southerly of which was on November 3 at 79° S latitude) and during the balloon phase (only) of eleven other flights. Also, good auroral soft radiation data were obtained on the rocket flight of November 3. Three satisfactory sets of magnetic field data on the equatorial electrojet were obtained on flights of up to 129 km in the vicinity of the Line Islands near the geomagnetic equator [Cahill 1959a]. In the southern auroral zone three magnetometer flights were conducted successfully (fig. 12).

Our work on the U.S.S. *Glacier* was secondary to the ship's principal mission as part of Operation Deep Freeze. Cahill and I were very grateful for the support of the captain, Cdr. B. J. Lauff, and especially that of the ship's operations' officer, Lt. (jg) Stephen O. Wilson, and Gunner's Mate (1/c) Gordon during many difficult operations at sea. Also, I wish to record my personal thanks to the ship's doctor, Lieutenant Commander Christy, who pulled me through a severe lymphatic infection that resulted from a minor leg injury.

From the Ross Sea the *Glacier* proceeded to Port Lyttleton, New Zealand. After closing out our shipboard operation, Cahill and I flew back with our precious data to the United States from Christchurch via Wellington, Auckland, the Fiji Islands, Canton, Hawaii, and San Francisco. We arrived in Iowa City on November 23, in time for Thanksgiving with our families. Cahill proceeded with the study of his magnetic field data and with preparation of a splendid Ph.D. dissertation. As is recounted in a later chapter, I was immediately caught up in a blizzard of activity preparing for the flight of *Explorer I* and still have a large body of data from our two 1957 expeditions that has not been fully reduced and published, even though I was quite familiar with the general run of the data during the course of the work.

In the context of the present account, the results of our 1957 Arctic and equatorial-Antarctic expeditions rounded out the epoch of scientific investigations precursory to the discovery of the radiation belts.

Courtesy of the U.S. Navy.

Fig. 12. An instrumented Loki II rocket being prepared for a rockoon flight from the USS Glacier *in the Antarctic (November 1957) by (left to right) Gunner's Mate (1/c) Gordon, Laurence J. Cahill, James A. Van Allen, and Lt. (jg) Stephen O. Wilson.*

IV.

Remarks on the Period 1946–57

The technical and scientific foundations of space exploration were firmly established during the period 1946–57 by observations through and above the earth's atmosphere using equipment carried on high-performance rockets. Indeed, a case can be made for adopting 1946 rather than 1957 as the start of the space age. The work during this important period was made possible by military rocket developments and large-scale logistic support by the U.S. Department of Defense. Moreover, almost all the scientific investigations themselves were supported by the military services, on an unclassified basis, with the enlightened point-of-view that prevailed following World War II and was perhaps best exemplified by that of the U.S. Office of Naval Research.

During this period U.S. investigators performed truly pioneering work from fixed launching sites, most notably the White Sands Proving Ground in New Mexico, and from ships at sea, ranging from Baffin Bay off the coast of Greenland to the Ross Sea in Antarctica. Many discoveries were made, and firm observational knowledge was obtained on matters that had previously been in the realm of educated conjecture. Among these were the structure of the earth's upper atmosphere and ionosphere, the ultraviolet and x-ray spectra of the sun, the distribution of ozone in the upper atmosphere, the intensity and latitude distribution of the primary cosmic radiation, the direct detection of energetic particles from the sun, the nature and intensity of the primary auroral radiation, the gravitational potential of the earth, and electrical currents in the ionosphere. Also, high-altitude photography of great portions of the surface of the earth and its cloud cover was performed and the foundations of global surveys of weather patterns and surface resources were laid [Holliday 1950]. The small cadre of investigators was motivated by the exploratory nature of the opportunities and by the great freedom and flexibility of the circumstances, notably free of long-range planning, detailed accountability, and other bureaucratic constraints. It was a period in which the wartime spirit—that most individuals are honorable and that results are what count—carried over into peacetime pursuits. From the perspective of 1983 one may deplore the passing of such an epoch. Those of us who survived this early period of high-altitude rocket investigations were, with few exceptions, the ones who conceived and built the first instruments for satellite flight. Without this preparatory period we would have been ill prepared to conduct investigations with satellites.

High-altitude research of a similar nature to that in the United States was also underway in the Soviet Union during the 1946–57 period, but because of political and language barriers the two bodies of work were effectively independent. Lesser but significant programs of rocket work were being conducted by French, English, Australian, and Canadian groups. Scientific programs utilizing rocket and satellite techniques became an integral part of the planning for the International Geophysical Year 1957–58. In turn, the IGY planning gave an important impetus to such programs and to related ones using more traditional techniques.

The great treatises of Chapman and Bartels [1940], Mitra [1948], Jánossy [1948], Alfvén [1950], Montgomery [1949], and Størmer [1955] were our constant companions as we planned future work.

V.

Plans for Scientific Work with Artificial Satellites of the Earth

The physical principles of placing a man-made object in orbit around the earth were clearly expounded by Isaac Newton in his justly celebrated treatise *Philosophiae Naturalis Principia Mathematica*, published in 1687. He also understood the basic principle of rocket propulsion. But during nearly three centuries thereafter no one succeeded in imparting the necessary velocity to a material object, a velocity calculated by Newton to be 7.9 kilometers per second. Pioneers in rocket propulsion in the late nineteenth and early twentieth centuries visualized how this might be done and, by bold engineering extrapolation, began to think that they could actually do it. The World War II development of high-performance rockets in Germany and to a lesser degree in the United States and the Soviet Union represented the first really powerful attack on this objective. Such developments were intensified in the United States and Soviet Union immediately after World War II. Popular magazines were filled with conjectures about space flight. By the late 1940s even hard-headed practitioners such as I began to be converted to the realistic prospect for earth-orbiting satellites and spacecraft that could escape from the earth's gravitational field. In July 1948 I wrote a paper entitled "The Use of Rockets in Upper Atmosphere Research" for the August 17–28, 1948, meeting of the Eighth General Assembly of the International Union of Geodesy and Geophysics in Oslo, Norway. After surveying the state of this subject, I concluded:

Then there is always the prospect of pioneering measurements at higher and ever higher altitudes. Serious consideration is being given to the development of a satellite missile which will continuously orbit around the earth at a distance of, say, 1000 km. In the even dimmer future is the prospect of astronomical type flights.

This thought was, of course, not original with me. It had been commonplace among visionaries for many years. Its inclusion in a working scientific paper by a working scientist in the field did, however, give it some significance. My statement was cited with levity by the *New Yorker* and with restrained contempt by the *New York Times*. The then director of the Applied Physics Laboratory, Ralph Gibson, asked me to strike it from the formally published paper on the grounds that it was excessively conjectural and detracted from the remainder of the paper. After some grumbling I complied.

Within a few years the flight of scientific equipment on satellites of the earth had been accepted as a realistic prospect throughout the interested segment of the scientific community.

At the forty-second meeting of the Upper Atmosphere Rocket Research Panel (UARRP) at the Ballistics Research Laboratories in Aberdeen, Maryland, on October 27, 1955, members Louis A. Delsasso and Homer E. Newell, in particular, argued that there was a timely need for a symposium on prospective scientific investigations with satellite-borne equipment. Both Delsasso and Newell were members of military laboratories and had full knowledge of the relevant plans for developing appropriate vehicles. According to the minutes of that meeting:

At the conclusion of the discussion, Dr. Delsasso moved that the UARRP proceed with plans for the symposium, that it include no classified material, that participation be limited to Panel members and such professional associates as they might wish to invite, and that the subject matter be confined to scientific aspects only. The motion was seconded by Dr. Newell and was carried unanimously.

It was further agreed as follows:

(a) The program will consist of formal papers of not to exceed twenty minutes duration each.
(b) Abstracts of 500 to 1000 words must be submitted in advance so that a printed assemblage of authors, titles and abstracts may be available at the time of the meeting.
(c) At least 100 copies of full papers must be made available by the authors at the time of the meeting.
(d) The papers should not assume advances in technology which are unlikely to be achieved within the next five years.
(e) *Admissible* topics are the following: (1) Proposed experiments and observations appropriate to small satellites (less than 100 pounds); (2) theoretical and interpretative considerations pertaining to such observations; and (3) techniques and components of novel or special applicability.
(f) *Inadmissible* topics are: (1) Rocket propulsion and guidance technology having to do with the vehicle except insofar as such matters influence the feasibility of the scientific uses, (2) space medicine, (3) political, public relations, and economic considerations.
(g) The symposium will be held at the University of Michigan at Ann Arbor on the 26th and 27th of January 1956.

(h) The chairman will proceed with assembling contributions, arranging the program, etc. Mr. Spencer of the University of Michigan will be in charge of local arrangements.
(i) It is intended that the contributions be published subsequently in collected form but no agreement was reached on exactly how to proceed on this matter.

The symposium was held as planned, being the forty-third meeting of the UARRP and a celebration of its tenth anniversary. A compilation of thirty-three papers given at the meeting was published promptly in a book I edited entitled *Scientific Uses of Earth Satellites* [1956]. The scope of the meeting is shown by the following preface and table of contents.

Preface

On 26, 27 January 1956, the Upper Atmosphere Rocket Research Panel held its tenth anniversary meeting at The University of Michigan in Ann Arbor. This meeting was devoted exclusively to specific, detailed proposals for the scientific use of small, artificial satellites of the earth. The present volume is a compilation of thirty-three of the papers which were presented there. The authors are, for the most part, seasoned veterans of physical research at high altitudes, using rockets as vehicles. Such work forms the tangible foundation for the competent utilization of satellites for scientific purposes.

The authors, the editor, and the publishers are well aware of the evanescent nature of descriptions of proposed experiments. Yet it has been felt worthwhile to make available in collected form the best early thinking on this subject. Much of the analytical work herein is of durable value. But the principal purpose of the book will have been served if it brings the potential value of artificial satellites to the attention of the scientific community at large and stimulates broad professional participation in the great and continuing undertaking of extending human knowledge of our physical environment by every conceivable means.

The Upper Atmosphere Rocket Research Panel, though an unofficial and informal body, has been the special custodian of the scientific integrity and the spirit of high-altitude research with rockets in the United States since the earliest beginnings of the subject in 1946. The present book is in effect Panel Report No. 43 of a long series of docu-

vi *Preface*

ments which reflect better than any other written record the heart-beat of this field of research. The present membership of the Panel is as follows: Warren W. Berning, Ballistic Research Laboratories; Louis A. Delsasso, Ballistic Research Laboratories; William G. Dow, University of Michigan; Charles F. Green, General Electric Company; Leslie M. Jones, University of Michigan; George K. Megerian, Executive Secretary, General Electric Company; Homer E. Newell, Jr., Naval Research Laboratory; Marcus D. O'Day, Air Force Cambridge Research Center; William H. Pickering, Jet Propulsion Laboratory; William G. Stroud, Signal Corps Engineering Laboratories; John W. Townsend, Naval Research Laboratory; James A. Van Allen, Chairman, State University of Iowa; Fred L. Whipple, Harvard College Observatory; and Peter H. Wyckoff, Air Force Cambridge Research Center.

JAMES A. VAN ALLEN
Editor

Iowa City, Iowa
1 May 1956

Contents

1. The Orbit of a Small Earth Satellite   1
R. J. Davis, F. L. Whipple, and J. B. Zirker

2. Time Available for the Optical Observation of an Earth Satellite   23
J. B. Zirker, F. L. Whipple, and R. J. Davis

3. Satellite Tracking by Electronic Optical Instrumentation   29
Harrison J. Merrill

4. Possibility of Visual Tracking of a Satellite   39
Donald E. Hudson

5. Interpretations of Observed Perturbations on a Minimal Earth Satellite   44
Jackson L. Sedwick, Jr.

6. Systems Design Considerations for Satellite Instrumentation   49
L. G. deBey

Contents

7. Components for Instrumentation of Satellites 55
H. K. Ziegler

8. Experiments for Measuring Temperature, Meteor Penetration, and Surface Erosion of a Satellite Vehicle 68
Herman E. LaGow

9. Insolation of the Upper Atmosphere and of a Satellite 73
P. R. Gast

10. Satellite Drag and Air-Density Measurements 85
L. M. Jones and F. L. Bartman

11. On the Determination of Air Density from a Satellite 99
Lyman Spitzer, Jr.

12. Pressure and Density Measurements Through Partial Pressures of Atmospheric Components at Minimum Satellite Altitudes 109
H. S. Sicinski, N. W. Spencer, and R. L. Boggess

13. Meteorological Measurements from a Satellite Vehicle 119
W. G. Stroud and W. Nordberg

14. The Radiative Heat Transfer of Planet Earth 133
Jean I. F. King

15. Visibility from a Satellite at High Altitudes 137
V. J. Stakutis and Capt. Joseph X. Brennan, USAF

16. A Lyman Alpha Experiment for the Vanguard Satellite 147
T. A. Chubb, H. Friedman, and J. Kupperian

17. A Satellite Experiment to Determine the Distribution of Hydrogen in Space 152
T. A. Chubb, H. Friedman, and J. Kupperian

Contents

18. Ultraviolet Stellar Magnitudes 157
Robert J. Davis

19. Quantitative Intensity Measurements in the Extreme Ultraviolet 166
H. E. Hinteregger

20. Cosmic-Ray Observations in Earth Satellites 171
James A. Van Allen

21. Study of the Arrival of Auroral Radiations 188
James A. Van Allen

22. Proposed Measurement of Solar Stream Protons 194
Willard H. Bennett

23. Exploring the Atmosphere with a Satellite-Borne Magnetometer 198
E. H. Vestine

24. Measurements of the Earth's Magnetic Field from a Satellite Vehicle 215
S. F. Singer

25. Satellite Geomagnetic Measurements 234
J. P. Heppner

26. Geomagnetic Information Potentially Available from a Satellite 247
Ludwig Katz

27. Ionospheric Structure as Determined by a Minimal Artificial Satellite 253
Warren W. Berning

28. Temperature and Electron-Density Measurements in the Ionosphere by a Langmuir Probe 263
Gunnar Hok, H. S. Sicinski, and N. W. Spencer

29. A Satellite Propagation Experiment   **268**
L. M. Hartman and R. P. Haviland

30. Electromagnetic Propagation Studies with a Satellite Vehicle   **276**
Fred B. Daniels

31. Study of Fine Structure and Irregularities of the Ionosphere with Rockets and Satellites   **283**
Wolfgang Pfister

32. Meteoric Bombardment   **292**
Maurice Dubin

33. Measurements of Interplanetary Dust   **301**
S. F. Singer

Even today, this book is good reading. The papers cover with remarkable completeness the program of scientific work that was actually conducted with satellites for many years thereafter. The principal distinction of most of the papers is that they represented realistic plans by seasoned practitioners of high-altitude research with rocket vehicles. Many of the authors became central figures in the subsequent space program of the United States. Such was the quality of this occasion.

Meanwhile, plans for the International Geophysical Year were getting underway and a satellite program was being developed within the United States as well as the Soviet Union. In early October 1955 the chairman of the U.S. National Committee for the IGY, Joseph Kaplan, established a Technical Panel on the Earth Satellite Program (TPESP) with Richard Porter as chairman and Hugh Odishaw as executive secretary. Other members in addition to Kaplan were Homer E. Newell, William H. Pickering, Athelstan F. Spilhaus, Lyman Spitzer, Fred L. Whipple, and myself. We met first on the October 20, 1955. In late January 1956 Porter asked me to serve as chairman of the Working Group on Internal Instrumentation (i.e., scientific instruments to be carried on the satellites), or WGII, and Pickering to serve as chairman of the Working Group on External Instrumentation (e.g., ground station tracking and telemetry). Other members of the WGII were Leroy Alldredge, Michael Ference, William Kellogg, and Herbert Friedman, as well as Spitzer, Porter, and Odishaw.

The principal functions of the WGII were the screening of proposals for investigations involving flight instruments, the establishment of flight priorities, and the consideration of telemetry, environmental testing procedures, budgets, etc. At the first meeting on March 2, 1956, we adopted and applied the following criteria for the assessment of proposed investigations:

(a) *Scientific Importance.* This aspect was taken to be measured by the extent to which the proposed observations, if successful, would contribute to the clarification and understanding of large bodies of phenomena and/or by the extent to which the proposed observations would be likely to lead to the discovery of new phenomena.

(b) *Technical Feasibility.* This criterion encompassed evidence for previous successful use of the proposed technique in rockets (or otherwise), apparent adaptability of the instrumentation to the physical conditions and data transmission potentialities of presently planned satellites, nature of data to be expected, and feasibility of interpretation of observations into fundamental data.

(c) *Competence.* An assessment of competence of persons and agencies making proposals was attempted. The principal foundation for such assessment was previous record of achievement in work of the general nature proposed.

(*d*) *Importance of a Satellite Vehicle to Proposed Work.* The nature of each proposal was analyzed with respect to the questions: Is a satellite essential or very strongly desirable as a vehicle for the observing equipment proposed? Or could the observations be made nearly as well or better with balloons or conventional rockets as vehicles?

At its second meeting on June 1, 1956, the WGII recommended that four of the twenty-five proposed investigations be "assigned Flight Priority A—with the understanding that if these developments proceed satisfactorily, the first 'few' vehicles will be assigned to their use." These were:

ESP-8   "Satellite Environmental Measurements"—H. E. LaGow, Naval Research Laboratory

ESP-9   "Solar Lyman-Alpha Intensity"—H. Friedman, Naval Research Laboratory

ESP-11 "Proposal for Cosmic Ray Observations in Earth Satellites"—J. A. Van Allen, State University of Iowa

ESP-4   "Proposal for the Measurement of Interplanetary Matter from the Earth Satellite"—M. Dubin, Air Force Cambridge Research Center

Six other investigations were assigned Flight Priority B and were recommended for developmental funding. The Vanguard group at the Naval Research Laboratory had been previously established as the lead agency for the development of satellite structures, telemetry transmitters, command receivers, thermal design, antennas, environmental testing, and other basic support services in addition to its major responsibility for development of the Vanguard launching vehicle [Green and Lomask 1970]. Funding for the development of scientific instruments and for data analysis and interpretation was to be provided by the National Science Foundation for groups that were not supported by other federal agencies.

The WGII continued to work closely with the NRL group in defining technical specifications for flight instruments and many related matters and to report to the TPESP on the progress of instrumental developments.

By early 1957 development of the scientific instruments was proceeding well, but grave difficulties in development of the Vanguard vehicle to a state of reliability were becoming increasingly worrisome.

I will return to this matter in a later chapter.

VI.

Sputnik I

Meanwhile, momentous developments were transpiring elsewhere, developments that profoundly affected my professional future and that of my small research group at Iowa.

Cahill and I had departed on September 23, 1957, from Boston on the U.S.S. *Glacier* for our second IGY rockoon launching expedition (chapter 3). On October 4 the ship, then in the vicinity of the Galápagos Islands after transiting the Panama Canal and steaming southwesterly for the Antarctic, received news via the Armed Forces Radio that the Soviet Union had placed a large satellite in orbit around the earth. Our response to this news is recorded in my field notebook SUI No. 7, labeled "Equatorial-Antarctic Expedition," under the date of October 5, 1957. The following account contains some verbatim excerpts:

> Yesterday night the 4th and early this morning were very exciting for me (as well as for the civilized world in general).
>
> Just before dinner time Larry Cahill told me that news was just coming in on the ship's news circuit that the *Soviet Union had successfully launched a satellite*. Factual details as follows:
>
> Inclination of orbit 65° to earth's equator. Diameter 58 cm, Weight 83.6 kilogram (Wow!). Estimated Height 900 kilometers [Perigee or Apogee?] Period $1^h 35^m$.
> Transmitted Signal: 20.005 mc/sec and 40.00 (?) mc/sec with switching alternately from one to the other—spending about 0.3 sec on each frequency. Would pass over Moscow at 1:46 AM and at 6:22 AM on the 5th, Moscow time. (Moscow is -3 or rather $+21$ zone time from Greenwich.)
> Our Ship's Position ~5° 30' N 92° W
>
> [+6 zone time]
>
> After dinner (and a very poor movie) I went up to the communications shack to see if there was any further news available (about 2120 Ship's time (+6). As I walked in to look at the teletype machine a young radioman [David Armbrust RM 3/c] wearing a pair of earphones and hovering over one of the ship's communications receivers turned to me and said—"I

think that I have it!" This was at 2120 [zone time of October 4] or 0320 Z (Greenwich time) of the 5th of October. I listened to the phones and heard a repetitive Beep-Beep-Beep — etc. of an audio frequency tone— loud and clear. The r.f. frequency was very nearly 20.005 mc/sec. I had earlier considered using our Clarke receiver but recalled that 55 mc/sec was [its] lowest frequency. Then I briefly considered the ship's capabilities but (too hastily) discarded this possibility on the general impression that the signal would be quite weak ala U.S. plans and that the ship's communications gear would be inadequate in basic noise level.

However, Mr. John Gniewek (formerly B.A. in physics fromm Syracuse Univ.), young civilian employee of the U.S. Coast and Geodetic Survey, who was a passenger on the Glacier going to the Antarctic to operate a magnetometer station there for the coming year, had been up to the communications shack earlier and had inquired if they could receive it. Armbrust had started looking with first success at ~0320 Z. He had also run a receiver calibration and had been listening and searching assiduously for some minutes.

My first reaction was: Could it possibly be true that this *was* the satellite's transmission? (Not a spurious effect of some kind—or something from WWV at 20.000 mc/sec, etc.

At about this time Gniewek came up. He listened, also excitedly. It immediately occurred to me that we should make a recording! I thought of our Ampex in rockoon lab but was somewhat discouraged of hauling it up to the comm. shack because of its weight and the way in which it was "built-in" to Cahill's apparatus! I remarked on this to Gniewek! He immediately responded that he had a small magnetic tape recorder in his room which he could easily bring up. I said fine! and rushed down to our rockoon lab to bring up my small Tektronix (Type 310) oscilloscope to look at the signal visually. I first noted the time as 0329 Z on the clock in the comm. shack. Within about 5 minutes we were both in operation! I immediately found the following appearance on the scope. [fig. 13]

The first pass that we received lasted from 0320 to 0349 Z or about 29 minutes. This length of time was about twice the 15 minutes that I calculated for horizon-to-horizon passage using the announced orbital parameters. We recorded a second pass from 0455 to 0521 Z and a third pass from 0629 to 0650 Z. At this time we were joined by Lt. (jg) Steve Wilson (our special project officer), Cahill, and several of the ship's officers including the captain, Cdr. B. J. Lauff, USN. We had a lively discussion of validity. The r.f. frequency and nature of the signal and the interval of time between the centers of successive passes were all in agreement with the Soviet announcement. The principal puzzle was how we could be receiving the signal over a calculated arc of over 100° of the orbit. I suggested ionospheric refraction as the explanation. I was able to convince my colleagues of this by noting that we routinely received WWV signals from Arlington, Virginia, at 20.000 mc s^{-1} by reflection from the ionosphere, and hence that ionospheric refraction of the *Sputnik I* signals when the transmitter was far below the horizon was, at least, qualitatively rea-

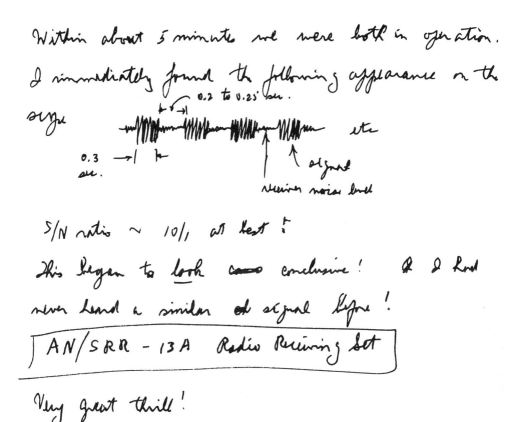

Courtesy of J. A. Van Allen.

Fig. 13. Copy of an October 5, 1957, entry in the author's field notebook recording the reception of the radio signal from Sputnik I.

sonable. Lauff suggested the Doppler shift of the r.f. frequency as the reason that the tuning of the receiver needed continuous adjusting during a pass and as a method to further validate that the signal was from a satellite. We made a crude calibration of the vernier dial on the receiver by listening to the beat note using a lab transmitter and found the apparent Doppler shift to be several hundred to a thousand cycles per second in rough agreement with the calculated value of $\pm\ 600$ c.p.s.

I then wrote the following dispatch which the captain released:

```
USS GLACIER                                    UNCLASS ROUTINE
IGY WASHINGTON DC
NR___ -T-R-050745Z - FM USS GLACIER - TO IGY WASHDC GR 60
BT . . . RECEIVED SIGNAL AT 20PT005 MC BELIEVED RUSSIAN
```

SATELLITE TRANSMITTER DURING FOLLOWING PERIODS 0320Z TO 0349Z AND 0455Z TO 0521Z AND 0629Z TO 0650Z X ON PERIODS 0PT25 TO 0PT30 SECOND DURATION CMM OFF PERIODS 0PT2 TO 0PT25 SECONDS X SHIPS RECEIVER AN/SRR-13A X POSITION 5-30N 92W X DISCOVERED BY DAVID ARMBRUST RM3 AND CONFIRMED AND RECORDED BY IOWA SCIENTISTS ABOARD . . . . . . BT
DR VAN ALLEN _____
MR GNIEWEK _____

OUTGOING
WU:JT 5 OCTOBER 57 050745Z

We received a fourth pass 0806 to 0823 Z with careful attention to the Doppler shift, and, after most of us had gone to bed, Gniewek received a fifth pass 0915 to 1010 Z.

On the following morning Gniewek played back one of his tape recordings to our Brush pen-and-ink recorder through a simple diode rectifier and low-pass RC filter (time constant 7 milliseconds) that I had assembled for the purpose. (A sample of this record is shown in figure 14.)

My summary thoughts on the situation were recorded in my field notebook on October 5 as follows (omitting some elementary calculations):

Items:
 1. Brilliant achievement!
 2. Tremendous propaganda coup for U.S.S.R.—also coming during CSAGI Rocket and Satellite Conference in Washington.
 3. Vehicle must be $\sim \frac{184}{22}$ = 9 times as heavy as Vanguard throughout if of same propulsive efficiency! $\sim$100 tons gross launching weight (also guidance accuracy!).
 4. Confirms my disgust with the Stewart Committee's decision to favor N.R.L over the Redstone proposal of Sept. 1955!!
 5. May lead to intensified U. S. Effort—Our five year plan!
 6. Causes me to be very sorry to miss the inevitable reconsideration and perhaps marked changes of the U. S. program. May be a genuine loss of opportunity for us at S.U.I. to assume a larger role in the future! By T.W.X. news today the CSAGI proposed the setting up of an international committee of not to exceed *six men* for the coordination of rocket and satellite programs!
 7. Very sensible choice of frequencies for ionospheric information and radio amateur interest.
 8. Evidently rather high power.
   Probably, Radiated Power $\sim$ or $>$ 10 watts.
   Perhaps as much as 100 watts.
 9. Assume 10 watts and 50% efficiency = 20 watts consumed. Assume, out of the total weight announced of 184 lbs. that 150 lbs. are batteries giving 40 watt-hr/lb.

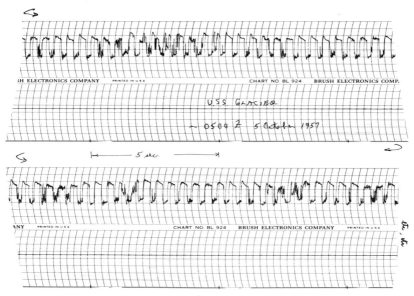

Courtesy of J. A. Van Allen.

Fig. 14. Copy of a paper-tape record of the rectified radio signal from Sputnik I *on October 5, 1957.*

      = 6000 watt hr.
      or 300 hours of operation
      or ~2 wks operation!
May have solar batteries, though there has been no hint of this in the announcements and I judge their inclusion to be somewhat out of character with the simple, brute-force approach of the Russians!
10. ?? Where do we stand now on Vanguard?
11. The decision ~2 months ago to "regroup" the Vanguard program and Townsend's hint in Cambridge of Security Council action = advance knowledge in U. S. of status of the Russian program!
12. The pompous character of the White House announcements!
13. The feebleness of the optical approach to tracking vs the radio approach strongly demonstrated as I have urged for some time.
14. "Our" highbrow choice of 108 mc/sec and very weak signal out of reach of most amateurs and the average population.
15. The astute choice by U.S.S.R. of frequencies which literally millions of persons can hear directly!
16. [Elementary orbital calculations]
17. [Calculations concerning ionospheric refraction of radio signals]
18. Would be of very great interest to know the observed duration of the 40 mc/sec signal!
19. Russians have a very great scientific lead on us in having a 65° inclination—for cosmic rays, aurorae, etc.

During the night of October 5–6 we received three more passes of *Sputnik I*. There were many news reports on the ship's radio from individuals who had received the radio signals at points around the world. The first authenticated visual observation was reported from Tasmania.

Cahill and I returned to our rockoon flight program with many thoughts but very meager information on the official U. S. reaction to *Sputnik I*. We were left to speculate on the consequences to our own professional futures.

On October 7, I was stricken with a severe case of lymphatic infection and a high fever resulting from a minor leg injury. I remained in bed under the care of Dr. Christy, the ship's doctor, and Dr. Slagle, a physician enroute to Little America. I was still immobilized, after heavy doses of penicillin and auremycin, as Cahill conducted the next rockoon flight on October 13. I was able to resume a regular working schedule on October 17.

VII.

Discovery of the Inner Radiation Belt of the Earth—Explorers I *and* III

In 1953 and 1954 I was on leave from the University of Iowa for fifteen months at Princeton University working on the design, construction, and operation of the Model B-1 stellarator, an early version of a laboratory device conceived by Lyman Spitzer for the magnetic confinement of ionized gas (plasma). This project was part of the then embryonic national effort to develop controlled nuclear fusion of deuterium or tritium or both as a potential source of electrical power. I was able to achieve confinement times of the order of a few milliseconds with Model B-1. Also, I became well grounded in the elementary principles of the motion of electrically charged particles in magnetic fields.

In 1954 Ernst Stuhlinger of the Army Ballistic Missile Agency (ABMA) in Huntsville, Alabama, called on me at my home in Princeton, New Jersey. Stuhlinger was a German physicist who had come to the United States in 1945 with Wernher von Braun as part of the Peenemünde V-2 group (cf. chapter 2). He had been especially helpful as an ombudsman for the visiting scientists at the White Sands Proving Ground in their dealings with the technical and engineering groups there and as a frequent participant in UARRP meetings. Stuhlinger's 1954 message was simple and eloquent. By virtue of ballistic missile developments at ABMA, it was realistic to expect that within a year or two a small scientific satellite could be propelled into a durable orbit around the earth. The ABMA plan was to use a multistage combination of military rockets with the liquid-fueled Redstone as the first stage. I expressed a keen interest in performing a worldwide survey of the cosmic-ray intensity above the atmosphere. This aspiration was one of long standing; a partial fulfillment of it was represented by the latitude survey with small rockets that I was engaged in at that time (chapters 2 and 3). It was obvious that even a few days of satellite observations would be equivalent to years of work by the rocket technique. Also, the necessary instrumentation was compatible with the limited weight, electrical power, and telemetry capacity envisioned by von Braun (the director of ABMA) and Stuhlinger for early satellites. Stuhlinger himself was interested in such cosmic-ray measurements, a matter that we had often discussed during our earlier acquaintanceship at the White Sands Proving Ground.

The earliest documented evidence that I have found of my specific planning for an instrument to be carried on an earth satellite is an internal University of Iowa four-page memorandum "Outline of a Proposed Cosmic Ray Experiment for Use in a Satellite (Preliminary)," which I wrote on November 1, 1954. The following are excerpts from that memorandum:

1. *Purpose of Experiment*

    (a) To measure total cosmic ray intensity above the atmosphere as a function of geomagnetic latitude and
    (b) To measure fluctuations in such intensity and their correlation with solar activity.

2. *Apparatus*

    Single Geiger counter, necessary auxiliary circuits, radio telemetering transmitter and antenna.

The subsequent sections summarized some preliminary technical considerations, viz.: 3. Choice of Power Supply; 4. Weight Analysis; 5. Fuel Supply; 6. Summary; 7. Solar Power; and 8. Comments. The naivete of the technical discussion is evidenced by the facts that I favored a hot gas (nitric acid and aniline) system with a gas turbine and rotating electrical generator as a power source (dead weight 9 lbs. and fuel weight 34 lbs. for one week's flight operation), that I estimated that the telemetering transmitter must radiate $\sim$ 5 watts for reception from a 1,500 mile line-of-sight distance, and that I expressed hope for solar batteries but remarked on the paucity of firm information on their practicality and availability. At the end of the memorandum I noted that "An apparatus for measurement of solar ultraviolet radiation would weigh about the same and require about the same power."

Following my 1954 meeting with Stuhlinger and my return to Iowa, I kept in touch with ABMA but was only vaguely aware of the intense rivalry among various segments of the Department of Defense (DOD) on the selection of a propulsion system for delivering satellites into orbit [Green and Lomask 1970]. On September 9, 1955, the deputy secretary of defense established Project Vanguard as the chosen alternative, thereby setting in motion a wide array of intensive preparations—in development of the three-stage vehicle itself as well as the launch facilities, a worldwide network of telemetry and tracking stations, payload structures, and many other supporting activities. Kurt Stehling of the Naval Research Laboratory, which had been selected as the agency of DOD to manage the Vanguard program, called me ten days later to fill me in on the situation. Anticipating a fast-breaking situation, I immediately prepared "A Proposal for Cosmic Ray Observations in Earth Satellites" and transmitted it on September 28, 1955, with a covering letter to Joseph Kaplan, chairman of the U.S. Committee for the International Geophysical Year (appendix A). On October 3 George Ludwig returned to Iowa City from our 1955 rockoon expedition to the Arctic, conducted by Frank McDonald, Joseph Kasper, Carl McIlwain, and him. Ludwig and I began preliminary conversations on how to actually build the instrument that I had proposed, and Ludwig addressed himself to learning how to make miniature, low-power electronic circuits using transistors, a then new device. In all of our previous rocket-borne instrumentation, we had used vacuum tubes. The transistor arrived on the technical scene just in time for the development of compact, rugged, reliable, and, above all, low-power drain circuitry for use in satellites. It made scientific research in space a realistic possibility in the face of very severe restraints on the weight and electrical power of instruments.

During December 1955 and January 1956 I prepared two detailed papers discussing the scientific rationale and some practical aspects of satellite observations of cosmic rays and auroral radiations and delivered these papers at the January 26–27, 1956, meeting of the UARRP (chapter 5). They were published as chapters 20 and 21 of *Scientific Uses of Earth Satellites* [Van Allen 1956] and were also used in a follow-on proposal to the U. S. National Committee for the IGY. The abstracts of these two papers are as follows:

Cosmic-Ray Observations in Earth Satellites
by James A. Van Allen
State University of Iowa

Abstract

Part A. Geographical Dependence and Temporal Variations of Cosmic-Ray Intensity in the Vicinity of the Earth

A single Geiger tube or scintillator carried in a satellite will make possible the study of the cosmic-ray intensity above the atmosphere on comprehensive geographical and temporal bases for the first time. The interpretation of expected data is outlined with respect to the following: determination of the effective geomagnetic field; the magnetic rigidity spectrum of the primary radiation; time variations of intensity and their correlations with solar and magnetic observations and with the observed intensity of secondaries observed in ground stations; and cosmic-ray albedo of the atmosphere. The monitoring function of a satellite will be especially valuable for providing an understanding of the extensive ground observations planned for the IGY period. The ideal data-transmission system and several practical compromises are discussed with regard to their consequences on the fullness of the data.

Part B. Relative Abundance of Heavy Nuclei in the Primary Cosmic Radiation

Satellite-borne instrumentation is uniquely able to solve one of the outstanding problems of the astrophysical nature of the primary cosmic radiation—namely, the abundances of Li, Be, and B nuclei. It is proposed to use a Cerenkov detector for this purpose. The technique has been developed and used intensively by this laboratory in balloon experiments. . . . Balloon apparatus suffers from the serious handicap of the residual atmosphere overhead; conventional rockets spend much too little time above the appreciable atmosphere. It appears to be feasible to fly a simplified version of this type apparatus in a satellite. Data can be accumulated for a complete revolution and played back in a 10-sec interval centered around passage over a meridian chain of telemetering receiving stations. The initial objective will be to measure the ratio of the sum of intensities of Li, Be, and B nuclei to the sum of the intensities of all heavier nuclei.

Study of the Arrival of Auroral Radiations
by James A. Van Allen
State University of Iowa

Abstract

The soft radiation discovered by the Iowa group above 50 km at auroral latitudes in 1953 . . . and further studied during 1954 and 1955 provides a novel foundation for plotting out the auroral zone by direct observation. A satellite in a nearly pole-to-pole orbit is a splendid vehicle for this purpose. If such orbits are not available in the near future, a good start on the problem can be made by means of balloon-launched, two-stage, solid-fuel rockets, fired along a meridian over a 10° span of latitudes including the auroral zone.

In the spring of 1956 Ludwig and I began work on specific detector systems and supporting circuitry for an instrument suitable for flight on a satellite. By this time we were beginning to understand the probable weight, power, size, and telemetry restraints of early U.S. satellites. Also, it appeared likely that the propulsive capability of either the Vanguard or ABMA system would restrict early launchings to a nearly due east direction from Cape Canaveral so that orbits would be limited to the approximate latitude range of 30° N to 30° S. For this reason we temporarily abandoned plans for auroral studies with satellite equipment, though we, of course, continued preparations for IGY rockoon expeditions to the Arctic and Antarctic.

We were acutely aware of the necessity for building satellite instruments to rigorous standards of reliability—i.e., insensitivity to temperature over a wide range, immunity to corona discharge in partial vacuum, generous tolerances on all operating elements, and mechanical ruggedness to resist the vibrations and linear and angular accelerations of launch. Our rocket experience was directly relevant, but we realized that we must play "hard ball" if we were to succeed in satellite work. Other requirements were minimizing the electrical power requirements and the weight and volume of the equipment.

All of these considerations led us in the direction of simplicity and meticulous attention to each element of the instrumentation. I settled on a single Geiger-Mueller tube as the basic radiation detector and obtained and tested samples of the halogen-quenched tubes developed by Nicholas Anton in his laboratory in New York. These tubes were mechanically rugged, had "infinite" lifetime, gave large signals, and operated stably over a wide range of temperatures—characteristics critically important for reliable operation on a satellite. Ludwig developed all of the circuits, including high-voltage and low-voltage power supplies, pulse amplifiers, scaling circuits, mixers, and modulators.

We also realized that real-time telemetry reception would be rather meager early in the program, whereas the purposes of our investigation demanded the fullest possible latitude, longitude, and altitude coverage and hence the addition of some form of data storage and playback, including a command receiver. In a progress report of May 30, 1956 [Ludwig and Van Allen 1956], we outlined the possible use of a magnetic drum recorder rotated stepwise once-a-second by a ratchet actuated by the scaled output of an electrically driven tuning fork. Later, Ludwig changed the storage element to a miniature magnetic tape recorder using metallic tape. The scheme contemplated the recording of GM tube pulses (scaled down by a factor not yet chosen at that time) over at least one complete orbit and command playback of the stored data at a high rate in the vicinity of a ground receiving station. The tuning fork also established an accurate time base on the tape. But we also included real-time data transmission as a back-up and supplemental mode.

On May 22, 1956, the sum of $106,375 for our satellite instrumentation program (ESP 32.1) was approved by the Technical Panel on the Earth Satellite Program, and we received initial funding from the National Science Foundation soon thereafter. The total authorized funding was later increased to $169,225.

In parallel with research, teaching, and technical preparations of our own instrument, I continued to be actively engaged in an intensive program of advisory activities and lectures on the potential uses of satellites for scientific work. In all of this I was, of course, motivated by direct interest as a participant, but I also considered that I was contributing to the development of long-range plans for a sustained program of satellite launchings. For example, on December 7, 1955, I lectured in New York City on "Experimental Uses of Earth Satellites" to about six hundred members of the local section of the American Rocket Society, the Institute of Electrical Engineers, and the Institute of Radio Engineers [Klass 1955]; on December 29 I spoke to the IGY symposium of the American Association for the Advancement of Science in Atlanta; on August 16, 1956, I gave four lectures on "Scientific Uses of High Altitude Rockets and Earth Satellites" at an MIT conference on "Orbital and Satellite Vehicles," arranged by Paul E. Sandorff; and during the period between September 9 and 13 at the Comité Spéciale de l'Année Géophysique Internationale (CSAGI) meeting in Barcelona, I gave a paper on the same subject and served as chairman of the Working Group on Rockets and Satellites. This international group comprised Russian, Chinese, French, British, Australian, Japanese, and Swiss, as well as U.S., scientists. One of the products of the working group was a set of resolutions dealing with international cooperation and exchange of data.

On September 20, 1956, the ABMA fired a three-stage rocket from the Patrick Air Force Base at Cape Canaveral, Florida, to an altitude of 1,050 km and a range of 5,300 km. An enlarged tank Redstone was used as the first stage, a bundle of eleven scale Sargeant motors as the second stage, and three scale Sargeant motors as the third stage. The "payload" on this test flight was an inert mock-up of a potential fourth stage. In the full ABMA concept of a four-stage satellite launching vehicle called Jupiter C, the fourth stage was to be a live scale Sargeant motor. A direct order from the chief of Army ordnance to von Braun prohibited the use of a live fourth stage on the test flight in order to preclude the attainment of orbital velocity. The reasons for this order were apparently military-political in nature—to avoid revealing the propulsive capability of the United States and to avoid alarming foreign nations with the realization that a U.S. satellite was flying over their territories. If a live fourth stage motor had been used and had burned properly, the first artificial satellite, with a useful payload of 9 kg, would have been placed in an orbit with a perigee altitude of about 300 km; and this would have been accomplished more than a year before the launching of *Sputnik I*.

On November 16 Stuhlinger advised me of these developments. He also expressed grave doubts about the realism of the Vanguard vehicle development on the promised time scale and contrasted that situation with the essentially flightworthy status of Jupiter C. In this context he encouraged me to suggest a specific cosmic-ray instrument that could be used for the purpose of payload design. I did so informally during this long telephone conversation, and, on the strength of Stuhlinger's judgment of the overall situation, I advocated the use of Jupiter C for the IGY program at the next meeting of the TPESP. Also, I decided that we would design the Iowa instrument in such a way that it would be suitable for the payload of either a Jupiter C or a Vanguard. Ludwig sent a detailed description of our instrument to ABMA in early February 1957. On April 19, 1957, Ernst Stuhlinger, Arthur Thompson, Charles Lundquist, and Joseph Boehn all of ABMA visited Ludwig, McDonald, and me at the University of Iowa and gave us a full description of the technical specifications of a satellite payload for the Jupiter C. These specifications enabled us to do detailed design work to assure compatibility of our instrument. By September 1957 Ludwig had completed construction and testing of our full package, nominally for the Vanguard but readily adaptable to the Jupiter C. The only unsolved aspect of the latter adaptation was the question of proper operability of the tape recorder in the rapidly spinning fourth stage of the Jupiter C.

Between July and November 1957 I was principally engaged in our Arctic and equatorial-Antarctic expeditions (chapter 3).

Following the successful launching of *Sputnik I* (chapter 6), as I learned later, von Braun, Pickering, and their associates at ABMA and JPL had been extremely active in developing a U.S. "response." In the course of these discussions Pickering pointed out that the Iowa cosmic-ray instrument was the only prime IGY instrument that had been configured for the Jupiter C payload, as an alternative to Vanguard. Von Braun, who had endorsed and fostered this decision, replied with mock innocence, "Isn't that interesting?" Following arrangements by Eberhardt Rechtin of JPL, Henry Richter and two others from JPL visited the University of Iowa on October 23 and reviewed the details of our instrument with Ludwig.

Rechtin met with Secretary of the Army Wilber M. Brucker and with a Mr. Holaday (DOD missile coordinator) on October 28 and received their approval of Jupiter C as a back-up satellite launcher. On October 30 I received the following radiogram on the USS *Glacier*:

> To Dr. Van Allen, Would you approve transfer of your experiment to us with two copies in spring. Please advise immediately.
>
> Pickering

On the same day and despite a certain lack of understanding of the significance of the message and a lack of firm knowledge of the approval of the Jupiter C, I replied as follows:

> Dr. W. H. Pickering
> Unable interpret your word transfer due ignorance recent developments. Our apparatus for original vehicle nearly finished. Delighted prepare three sets non-storage type for JPL program.
>
> Van Allen

On or about the same date Pickering had discussed his plans with Richard Porter, chairman of the TPESP, and requested his advice on a suitable payload. Porter recommended the following priority list:

1. Iowa Cosmic Ray Detector
2. NRL Environmental Measurement Package
3. NACA Inflatable Sphere

On November 2, I received the following further message from Pickering:

> Present planning suggests [flying] most of existing equipment but transferring responsibility to us instead of carrying two programs. First experi-

ment would be continuous. Second would be storage type. Suggest you phone me on arrival in New Zealand.

Upon arrival on the USS *Glacier* in Port Lyttleton, New Zealand, I received a third message from Pickering:

> Urgently need your approval on proposed change of Ludwig experiment. Porter committee has given their approval to proposal to change experiment to JPL and to modify experiments as agreed upon between Ludwig and JPL.

On November 13, still with some uneasiness about unqualified agreement, I sent a commercial cable to George Ludwig:

> Question. Is Pickering plan for our experiment agreeable with you? Please cable answer IGY rep Christchurch.

On the following day I received a response from Ernest Ray, another of my colleagues at Iowa:

> After high level approval and obvious rearrangement of old program, George left town for extended stay. George quite happy Pickering plans. Hope you say yes.

After receiving this assurance, I immediately advised Pickering as follows:

> Approve transfer our experiment accordance JPL plan.

Ludwig packed all of his technical equipment and payload components (and his family), cancelled his university registration, and moved between November 18 and 20 to Pasadena "for the duration." There he worked with the JPL staff in the detailed adaptation of our instrument to the Jupiter C payload [Richter et al. 1959]. The entire arrangement was called "Deal I." Because of the high rate of spin of the Jupiter C final stage, as compared with that of Vanguard, Ludwig omitted his magnetic tape data storage unit from Deal I; as a result we were to be exclusively dependent on real-time telemetry on the first flight. But he immediately undertook modifications of the tape recorder to reduce its susceptibility to centrifugal force and arranged for its on-axis positioning in the payload. Our plan for the tape recorder was that we could store the data from our detector over a complete orbit, then play it back on command within six seconds over a chosen receiving station. This configuration was called "Deal II." As of early December 1957 the planned launch readiness date was February 1, 1958, for Deal I and March 1, 1958, for Deal II. The date for Deal I, in particular, was compatible with von Braun's November 8 promise to President Dwight Eisenhower that the first Jupiter C with scientific payload would be ready to launch within ninety days. As mentioned above the reasonable nature of this promise had been demonstrated in September

1956 by the flight of Jupiter C (with inert upper stage) to a range of 5,300 km.

The completed payload for the first Jupiter C launch was delivered to Cape Canaveral in mid-January 1958 and mated to the upper stage (figs. 15, 16, 17). Ludwig accompanied the JPL group to the Cape for preflight checkout work and for the launch.

The Deal I (*Explorer I*) payload was lifted off the launch pad at Cape Canaveral 10:48 P.M. EST on January 31, 1958 (03:48 GMT of February 1). The vehicle consisted of four propulsive stages. The first stage was an upgraded ABMA Redstone liquid-fueled rocket. The second stage consisted of a cluster of eleven scale Sergeant motors of JPL development; the third, a cluster of three such motors; and the fourth and final stage, a single such motor to which the payload (instrumentation 4.82 kg, total orbital body 13.97 kg) was permanently attached. The burning of all four stages was monitored by down-range stations and judged to be nominal. The final burnout velocity of the fourth stage was somewhat higher than intended, and there was a significant uncertainty in the final direction of motion. Hence, the achievement of an orbit could not be established with

Courtesy of George H. Ludwig.

Fig. 15. Photograph of the scientific payload of Explorer I *with outer shell and nose cone removed [Ludwig 1959].*

Fig. 16. Drawing-to-scale of the principal elements of Explorer I. *As shown, the casing of the expended fourth-stage rocket remained attached to the scientific payload in flight [Ludwig 1959].*

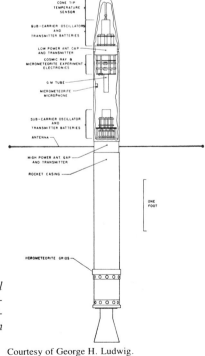

Courtesy of George H. Ludwig.

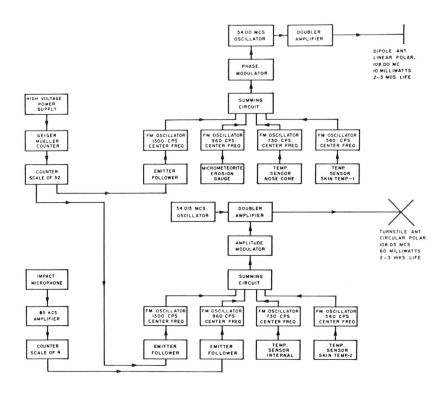

Courtesy of George H. Ludwig.

Fig. 17. Functional schematic of the electronics of Explorer I. *Electrical power for all circuits was derived from primary mercury batteries [Ludwig 1959].*

confidence from the available data. The telemetry transmitter was operating properly, and the counting rate data from our radiation instrument corresponded to expectations (fig. 18). The reception of the telemetry signal after the lapse of one orbit was necessary before success could be confirmed. The nominal period of the orbit was ninety-five minutes, and the first pass from west to east over northern Mexico was expected to provide the first clear opportunity for reception of the signal by stations in southern California.

By previous arrangement I was a member of a group in the War Room of the Pentagon, which served as a center of communications. Others present included Wernher von Braun, Secretary of the Army Wilber M. Brucker, General Lyman L. Lemnitzer, General John B. Medaris, and William H. Pickering. For about an hour following receipt of the down-

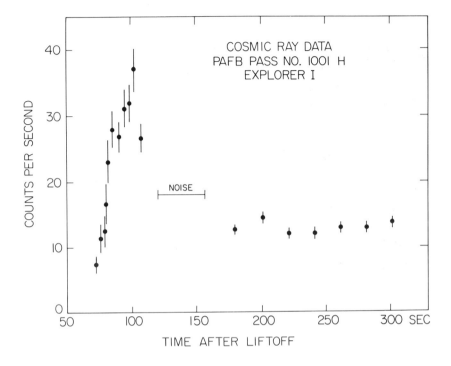

Courtesy of J. A. Van Allen.

Fig. 18. Counting rate of Geiger-Mueller tube on Explorer I *as a function of time after lift-off from Cape Canaveral on February 1, 1958.*

range station reports, there was an exasperating absence of information. Then there began a trickle of affirmative, amateur reports from around the world, none of which withstood critical scrutiny. The clock ticked away, and we all drank coffee to allay our collective anxiety. After some ninety minutes, all conversation ceased, and an air of dazed disappointment settled over the room. Then, nearly two hours after launch, a telephone report of confirmed reception of the radio signal by two professional stations in Earthquake Valley, California, was received. The roomful of people exploded with exhultation, and everyone was pounding each other on the back with mutual congratulations. Pickering, von Braun, and I were whisked by an army car from the Pentagon to the National Academy of Sciences and smuggled through a back door, where we made our preliminary report to Porter and the IGY staff. We were next led into the Great Hall of the

Courtesy of J. A. Van Allen.

Fig. 19. (Left to right) William H. Pickering, James A. Van Allen, and Wernher von Braun hold aloft a full-scale model of Explorer I *at a press conference in the Great Hall of the National Academy of Sciences in the early hours of February 1, 1958, following confirmation that* Explorer I *had completed its first orbit of the earth.*

Academy (by then about 1:30 A.M.) to report to the press. To my astonishment the room was nearly filled with reporters, photographers, and many other interested persons who had been waiting there since about 10:00 P.M. The ensuing press conference was a spirited one (fig. 19). The successful launch of *Explorer I* was an event of major national and international interest, coming as it did after three humiliating launch failures of Vanguard.

It was not until many weeks later that we were able to assemble a sufficiently comprehensive body of data to achieve a clear understanding of the scientific results from *Explorer I*. Meanwhile, we all went back to work preparing for the scheduled launches of two Deal II payloads. Ludwig returned to Pasadena, and I returned to Iowa City, where Ernest Ray and I started organizing the initially rather meager flow of data from our cosmic ray instrument. Also, I resumed my normal university duties. In a brief memorandum to the IGY staff and committees on February 28 [Van Allen and Ray 1958], we reported:

1. *General Remarks*

(a) The entire set of cosmic ray equipment including both high-power and low-power transmitters has apparently operated properly as follows: before launching, throughout the launching phase, and throughout the flight period 1 February to the early morning of 11 February.

(b) Records thus far received for passes since 11 February do not contain readable records of cosmic ray intensity

(c) During 12 February the high-power transmitter quit (or at least diminished in power output below the level for reception). On 23 February the high-power transmitter was again received and has continued to transmit up to the date of writing. Total battery life was estimated to be somewhat over two weeks.

(d) The low-power transmitter has operated continuously throughout the period 1 February to date. Its batteries are expected to supply adequate power until about 1 April.

. . . .

3. *Summary of Data*

. . . .

(c) [Data from 34 brief passes, mainly from stations in California, show that:] Rates of the Geiger counter lie within the range 12 to 80 counts per second corresponding to charged particle directional intensities J of about 0.12 to 0.80 particles/cm^2 sec steradian. All of these intensities are within the range of expected values (See Chapter 20 of "Scientific Uses of Earth Satellites," 1956) for the latitudes and altitudes being covered. . . .

Carl McIlwain, an advanced graduate student, was principally engaged at this time in preparing auroral particle detection instruments for rocket flight from Fort Churchill, but he also joined in the examination of *Explorer I* data. During February McIlwain conducted two Nike-Cajun rocket flights into visible aurorae and established that the "major fraction of the auroral light was produced by electrons with energies of less than 10 keV," thus confirming and greatly improving the inferences that we had made from our soft auroral radiation observations during the preceding several years [McIlwain 1960b].

Explorer II with our full-up payload (Deal II) including the magnetic tape recorder was launched on March 5, but an orbit was not achieved because the fourth-stage ignition failed.

On March 11 and 12 Ludwig and I met in Pasadena with Pickering, Jack Froelich, and Henry Richter of JPL and Wolfgang Panofsky of Stanford to review the radiation intensity data from *Explorer I*. We also discussed techniques for building miniature detectors that were suitable for small satellites but capable of particle identification and the measurement of energy spectra and angular distributions—characteristics not possessed by the single Geiger-Mueller tube that I had adopted for the early measurements. There were vague allusions at this meeting to the possibility of radiation experiments at high altitudes by the Atomic Energy Commission (AEC). It was some four weeks later before I learned that a series of high-altitude bursts of small nuclear bombs was being planned as a test of an idea that Nicholas Christofilos of the Livermore Radiation Laboratory had proposed. He had visualized that energetic, electrically charged particles could be artificially injected into the earth's magnetic field and that they would then be trapped therein as in a laboratory magnetic mirror machine. In fact, he was engaged in developing a dipolar magnetic field machine, called the Astron, for demonstrating confinement of hot plasma in the laboratory. There were also military implications of the possible production of high-intensity radiation regions around the earth and of the ionospheric effects of the initial bursts. Panofsky had been asked by the AEC to help assess the physical effects to be expected. Pickering had suggested that the Iowa group would be a suitable one for observing the effects. The proposed tests were classified as secret at that time. Pickering and Panofsky had developed the plan that an unclassified program of satellite observations with an improved version of our *Explorer I* instrument might be placed within an IGY context as a logical follow-on to *Explorer I* but that such an instrument would also serve the classified purpose.

I was very eager to join in this enterprise purely as an extension of our observations with *Explorer I*, but I was also intrigued by my then vague perception of the other possibilities. Ludwig and I started to plan an improved system of detectors. We were soon joined by McIlwain, who contributed his very valuable experience in developing instrumentation for auroral rocket flights.

Explorer III, carrying our full Deal II instrument, was successfully orbited by the third Jupiter C on March 26, 1958 (fig. 20). The tape recorder (the first such device ever flown in a satellite) functioned beautifully in response to ground command and fulfilled our plan of providing complete orbital coverage of radiation intensity data.

The assembly of data from *Explorer I* was proceeding rather slowly. The long slender body equipped with four whip antennas had passed from its original axial spin mode into the minimum kinetic energy state of a flat spin about a transverse axis as deduced from the modulation of the received signal—an impressive and humiliating lesson in the elementary mechanics of a somewhat nonrigid body. This modulation produced periodic fade-outs of the signal. In addition, the r.f. signals (at 108 MHz) from the 60 milliwatt and the 10 milliwatt transmitters were quite weak, at the best, and there were many other technical problems in getting reliable, noise-free recordings at the receiving stations. The whole telemetry situation was, of course, undergoing its first shakedown under fully realistic conditions. The overall result was that during the first few weeks of *Explorer I*'s orbital flight, we had only a sparse set of data consisting of segments of the order of one minute's duration from many different and somewhat uncertain positions in latitude, longitude, and altitude. The Na-

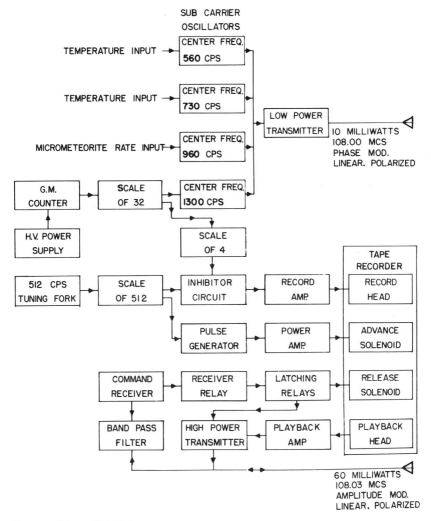

Courtesy of George H. Ludwig.

Fig. 20. Functional schematic of the electronics of Explorer III, *including the magnetic tape recorder and its associated circuitry [Ludwig 1961].*

val Research Laboratory tracking team under Joseph Siry's supervision was making a heroic effort to calculate a reliable orbit but was also having shakedown problems, partly because of the larger-than-expected eccentricity of the orbit.

During some segments of the data the counting rate of our single Geiger tube was of the order of 12 to 80 counts per second, generally within the range that we had expected for cosmic radiation, and the instrument appeared to be operating reliably. In other segments of data there were no counts observed for as long as two minutes. On one noteworthy pass the apparent rate underwent a transition from zero to a reasonable value within about twenty seconds. There was no conceivable way in which the cosmic-ray intensity could drop to zero at high altitudes. On the other hand we had a high level of confidence in the Geiger tube and the associated electronic circuitry because of its conservative design and the rigorous thermal and mechanical testing to which it had been subjected before flight. The puzzle hung over our heads as we tried to see if there was any systematic dependence of the apparent failure on passing through earth shadow, on payload temperature, on altitude, latitude or longitude. Noise-free data accumulated slowly. Also, the higher power transmitter failed after eleven days of flight. This premature failure spawned the hypothesis that a vital component had been disabled by a micrometeoroid hit, a possibility of neurotic concern at that time. In defiance of the hypothesis the transmitter resumed operation a few days later, but its operation thereafter was desultory. Throughout this early period we were heavily occupied, as indicated above, in preparing for the *Explorer II* and *III* launchings, in working out data reduction and analysis techniques (to which we had given relatively little attention before flight), in formulating plans for subsequent flights, and in coping with a steady flow of urgent phone calls on practical arrangements and on inquiries on our progress. Indeed, our original plan to accumulate a comprehensive body of data on the distribution of cosmic-ray intensity around the earth did not dictate urgency in analysis of data. We had high hopes for eventually getting a worldwide survey of the auroral soft radiation, but the decision to launch *Explorer I* nearly due east from Cape Canaveral in order to obtain the maximum advantage of the rotation of the earth yielded an orbital inclination of only 33.3° to the equator. Hence, the maximum geomagnetic latitude was only about 45°, far below the auroral zone.

Following the March 26 launching of *Explorer III*, I flew to Washington to confer with Siry, John Mengel, and others at the Naval Research Laboratory and to pick up preliminary data for *Explorer III*. Contrary to some popular accounts the Vanguard group fully supported the Explorer program in many vital ways. The first successful launch of Vanguard had occurred on March 17, and the NRL team was operating on an around-the-clock basis, coping now with tracking and data acquisition for three satellites. (I rode from downtown Washington to NRL in a taxi with an interesting young man named Charles Townes who gave me a brief education in masers enroute.) From NRL I returned to the Vanguard data reduction center on Pennsylvania Avenue and picked up the complete record of a successful playback of data from our *Explorer III* tape recorder. The playback had been received at the San Diego minitrack station on March 28. I put the record in my briefcase and returned to my hotel room, where, with the aid of graph paper, a ruler, and my slide rule, I worked out the counting rate of our Geiger tube as a function of time for a full 102-minute period and plotted the data (fig. 21).

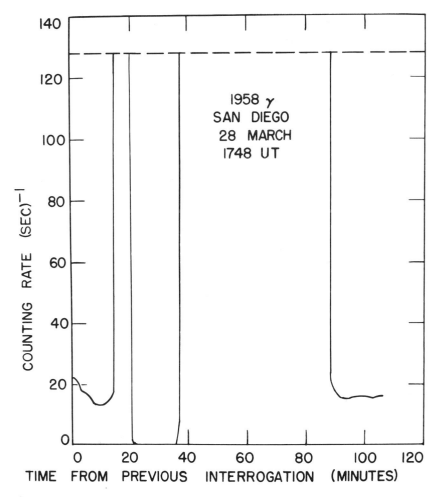

Courtesy of J. A. Van Allen.

Fig. 21. The first available full orbital set of radiation data from the tape recorder on Explorer III, *obtained from an interrogation over San Diego. The horizontal dashed line at 128 counts per second is the upper counting rate limit of the tape recorder system.*

The data provided a beautiful explication of the fragmentary information from *Explorer I*. The counting rate at low altitudes was in the expected range of 15 to 20 counts per second. There was then a very rapid increase to a rate exceeding 128 counts per second (the maximum recordable rate of our on-board storage system). A few minutes later, the rate decreased rapidly to zero. Then after about fifteen minutes, it rose rapidly from zero to greater than 128 counts per second and remained high for forty-five minutes, then again decreased rapidly to 18 counts per second as the orbit around the earth was nearly completed. At 3:00 A.M. I packed my work sheets and graph and turned in for the night with the conviction that our instruments on both *Explorers I* and *III* were working properly, but that we were encountering a mysterious physical effect of a real nature. Early the following day, I flew back to Iowa City and proudly displayed my graph to Ernie Ray and Carl McIlwain. During the previous day McIlwain had made tests with our prototype Geiger tube and circuit using a small x-ray machine and demonstrated that a true rate exceeding about 25,000 counts per second would indeed result in an apparent telemetered rate of zero. The conclusion was then immediate — at higher altitudes the intensity was actually at least a thousand times as great as the intensity due to cosmic radiation. Ray's famous (though consciously inaccurate) remark summarized the situation, "My God, space is radioactive!" Our realization that there was actually a very high intensity of radiation at high altitudes rationalized our entire body of data.

George Ludwig returned from JPL to the University of Iowa on April 11, and the four of us worked feverishly in analyzing the data from *Explorers I* and *III* (by primitive hand reduction of pen-and-ink recordings) and organizing them on an altitude, latitude, and longitude basis. A crucial aspect of the data was the repetitive, systematic dependence of the GM tube's counting rate on these parameters. I promptly informed Porter, Odishaw, Newell, and Pickering of our results. The latter informed Panofsky, whose reaction was that the Soviets had beaten us to the punch in conducting a Christofilos-type test. Odishaw admonished me to make no public announcement of our findings, pending a formal IGY report, which he would schedule as soon as we had our results in order. We agreed on a May 1 date for the public report.

During mid-April we prepared graphs and a short written statement of our raw findings [Van Allen et al. 1958], and I mulled over the meaning of the results. I entertained two quite different lines of thought: (*a*) that we might be detecting high-energy x rays or γ rays, possibly from the sun, or (*b*) that the high intensity radiation might be akin to the auroral soft radiation that we had studied during the preceding several years with rockoon flights at high latitudes and most recently with McIlwain's rocket flights from Fort Churchill and that we had identified as being electrons having energies of the order of 10s of keV [McIlwain 1960b]. I quickly rejected hypothesis *a* on the conclusive grounds that "the effect" was present during both daylight and dark conditions, that it exhibited a strong latitude effect, and that the extremely sharp increase in intensity with increasing altitude was impossible for any type of electromagnetic or corpuscular radiation. Specifically, the rapid increase occurred within an altitude range of less than 100 kilometers at an altitude of the order of 1000 km; the decrease in atmospheric thickness within that increment of altitude was totally negligible compared to the some 1.5 g cm^{-2} of material in the nose cone and wall of the counter. I concluded that the effect had to be attributed to electrically charged particles, constrained by the earth's external magnetic field from reaching lower altitudes. By virtue of my familiarity with an early paper of Størmer [1907] and with magnetic field confinement of charged particles in the laboratory during my 1953–54 work building and operating an early version of a stellarator at Princeton, I further concluded that the causative particles were present in trapped orbits in the geomagnetic field, moving in spiral paths back and forth between the northern and southern hemispheres and drifting slowly around the earth. The intensity of such trapped particles would be gradually diminished at low altitudes by atmospheric absorption and scattering.

The foregoing account of observations and interpretation is essentially the one I gave in a joint session of the American Physical Society and the National Academy of Sciences in the latter's auditorium on May 1, 1958. Fortunately, a tape recording was made of my lecture and of the ensuing question-and-answer period, though I did not learn that until a year or more later. A written transcription of this tape provides a documented, published record of this lecture, complete with grammatical errors and colloquial language [Van Allen 1961].

I had adopted at that time the working hypothesis that the trapped radiation consisted of "electrons and likely protons, energies of the order of 100 keV and down, mean energies probably about 30 keV." In this vein of thought the response of our Geiger counter would be attributed to bremsstrahlung produced as the electrons bombarded the nose cone of the instrument. If this bremsstrahlung interpretation were correct, I estimated, an omnidirectional intensity of 10^8 to 10^9 (cm^2 sec)$^{-1}$ of 40 keV electrons would be required to account for the observed counting rate at altitudes of ~1500 km over the equator. However, in my May 1 lecture as well as in response to a question at the end, I emphasized that we had no definitive identification of particle species and that the particles might be penetrating protons or penetrating electrons. I did, however, regard protons and electrons of energies necessary to penetrate the Geiger tube directly, namely $E_p > 35$ MeV and $E_e > 3$ MeV, as unlikely in view of our auroral zone measurements with rocket-borne equipment. A few months later, we showed that this opinion was mistaken.

Of the some 1,500 real-time recordings [Ludwig 1959] of *Explorer I* telemetry signals in the period February 1 to May 9, 1958, 850 contained readable cosmic-ray data. The tape-recorded data from *Explorer III* had

Courtesy of the World Data Center, National Academy of Sciences.

Fig. 22. A sample of Explorer III *in-flight data on radiation intensity recorded on the magnetic tape recorder in the payload and later played back over the telemetry link upon radio command to a ground receiving station. The equally spaced pulses at one-second intervals are derived from a tuning-fork-driven oscillator. Missing (or inhibited) pulses correspond to the accumulation of 128 counts from the Geiger-Mueller tube. If the counting rate exceeds 128 counts per second, all pulses are missing. If the true counting rate exceeds about 25,000 counts per second, there are no missing pulses because the apparent rate is then near zero. The raw counting rate in this example is about 21 counts per second.*

an upper rate limit of 128 counts per second, a value that we had judged to be adequate for the originally planned cosmic-ray investigation (fig. 22). However, the real time transmissions from both *Explorers I* and *III* had a much greater dynamic range. For this mode the Geiger tube pulses were scaled by a factor of 32 and then applied as a two-level voltage signal to a voltage-controlled oscillator having a center frequency of 1,300 Hz. The discontinuously jumping frequency of the VCO was used to modulate the transmitter. The signal received and recorded at a ground station was eventually converted back to a square wave (of statistically varying width) and displayed as a function of time on the strip chart of a pen-and-ink recorder (fig. 23). Hence, the practical upper limit on the apparent counting rate of the Geiger tube was dependent on overall signal noise ratio. In favorable cases we were able to read scaled rates as high as 100 per second, corresponding to an apparent Geiger tube rate of 3,200 counts per second. Because of the dead time of the Geiger tube and associated circuitry (168 microseconds), the maximum apparent rate of the output was 2,200 counts per second. Finally, using the laboratory-measured relationship (fig. 24) of apparent counting rate to true counting rate (the rate that would have been observed if the dead time had been zero), we were able to extend the dynamic range to about 60,000 counts per second [Yoshida, Ludwig, and Van Allen 1960]. The measured counting volume of the GM

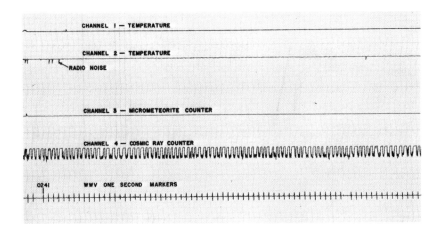

Courtesy of George H. Ludwig.

Fig. 23. Chart recording of the reduced telemetry signal from Explorer I, *Patrick Air Force Base, 0241 UT, February 4, 1958. Channel 4 shows the counting rate data from the Geiger-Mueller tube—about 40 counts per second after having been scaled down by a factor of 32 before transmission [Ludwig 1959].*

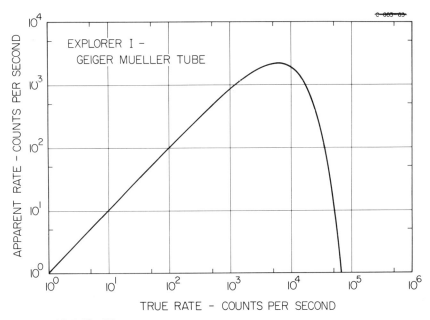

Courtesy of J. A. Van Allen.

Fig. 24. Relationship of the apparent (observed) counting rate to the true counting rate of the Geiger-Mueller tube on Explorer I.

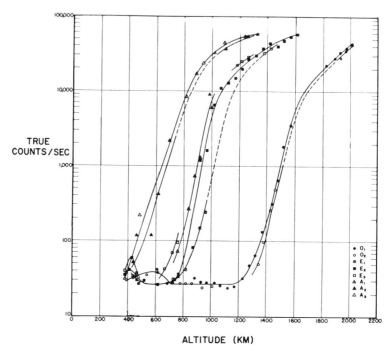

S. Yoshida, G. H. Ludwig, and J. A. Van Allen, "Distribution of Trapped Radiation in the Geomagnetic Field," *Journal of Geophysical Research*, vol. 65 (1960), p. 810, copyrighted by the American Geophysical Union.

Fig. 25. True counting rate (i.e., observed rate after correction for dead time) of the Geiger-Mueller tube on Explorer I *as a function of altitude over eight geographic regions, all near the magnetic dip equator. The two curves on the left are for regions in the approximate geographic longitude 290° E; the central bundle of curves, 5° E; and the two curves on the right, 105° E [Yoshida, Ludwig, and Van Allen 1960].*

tube (Type 314 of the Anton Electronics Laboratories) was cylindrical in shape with a diameter of 2.0 cm and an effective length of 10.2 cm so that the omnidirectional geometric factor was 17.4 cm^2. Thus, a counting rate of 60,000 counts per second corresponds to an omnidirectional intensity of penetrating particles of 3.4×10^3 (cm^2 sec)$^{-1}$.

A much larger body of data was obtained from *Explorer III*. During its forty-four days of useful life the satellite completed 523 orbits around the earth and 504 tape recorder playbacks were attempted. Of these attempts a total of 408, or 81%, were successful. The tape recorder continued to operate perfectly throughout this period and never failed to respond to a command that was electronically successful [Ludwig 1961].

Data reduction was continued during 1958 and 1959. Virtually all of the *Explorer I* data were reduced and analyzed [Yoshida, Ludwig, and Van Allen 1960; Loftus 1969] (figs. 25, 26). But even as of 1983, George

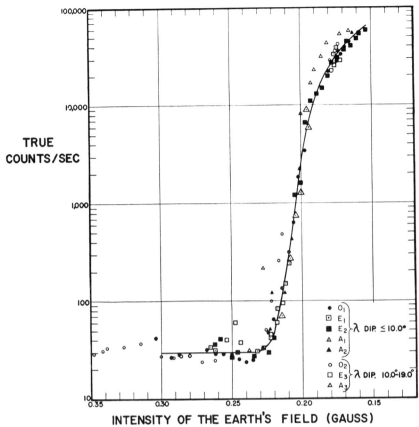

S. Yoshida, G. H. Ludwig, and J. A. Van Allen, "Distribution of Trapped Radiation in the Geomagnetic Field," *Journal of Geophysical Research*, vol. 65 (1960), p. 809, copyrighted by the American Geophysical Union.

Fig. 26. The counting rates of figure 25 replotted as a function of the intensity of the earth's magnetic field at the points of observation [Yoshida, Ludwig, and Van Allen 1960].

Ludwig and I regret not having completed the reduction of the major part of both the real-time and stored *Explorer III* data. In early 1958, however, we thought that we had gleaned the principal elements of the data and hardly felt it worthwhile to continue; the dynamic range of the tape recorded data was limited; and we then had the prospect of follow-on satellite missions with detectors specifically designed for the then-known intensities and for particle species identification and other more discriminating properties of the radiation.

A description of the instrumentation and the telemetry scheme of *Explorer I* had been made available through the International Geophysical Year organization prior to the launching, and the approximate elements of its orbit were announced soon after the launching. The choice of radio frequency (108 MHz) and the low radiated power (60 and 10 milliwatts) of the two transmitters, together with the unanticipated tumbling of the satellite, made reception of satisfactory telemetry signals difficult, as noted above. Nonetheless, Japanese workers at the Radio Research Laboratories at Kokubunji, Tokyo (35° 42' N, 139° 29' E), succeeded in receiving workable signals during the first eleven days of February 1958 within the altitude range 340 to 2,100 km. The observed data on the counting rate of the Geiger tube were reduced and published on a rapid time scale in two separate papers [Aono and Kawakami 1958; Miyazaki and Takeuchi 1958]. Indeed, the first of these was the earliest formally published paper on *Explorer I* data. The altitude dependence of the counting rate was determined. It showed, with some fluctuations, a gradual increase up to about 1,200 km and an exceedingly rapid increase (a factor of 3 per 150 km) above that altitude [cf. Yoshida, Ludwig, and Van Allen 1960]. Also, a significant increase of counting rate occurred at an altitude of 1,000 to 1,100 km during the great geomagnetic storm of February 11, a matter studied later and more comprehensively by Loftus [1969]. These data were of a similar nature to those available to the Iowa group during February and March 1958, but the Japanese workers gave no hint of interpreting the strong altitude dependence of counting rate above 1,200 km as evidence for a population of geomagnetically trapped energetic particles. However, it may be noted that the limited geographical coverage of the Japanese data set would have made such an interpretation difficult to sustain even if this line of thought had occurred to the investigators.

Later in 1958 Miyazaki and Tokeuchi [1960] recorded data from *Explorer IV* and compared them with those from *Explorer I*.

I am not aware of any other non–U.S. publication of data from *Explorers I, III*, or *IV*.

In our early reports, I used the term "geomagnetically trapped corpuscular radiation." At the press conference following the May 1, 1958, lectures at the National Academy of Sciences, I described the distribution of the radiation as encircling the earth. A reporter asked "Do you mean like a belt?" I replied: "Yes, like a belt." This was the origin of the term *radiation belt*. At a meeting sponsored by the International Atomic Energy Agency in Europe in the summer of 1958, Robert Jastrow first used the term *Van Allen radiation belt*.

The more inclusive term *magnetosphere* was suggested by Gold [1959a]: "It has now become possible to investigate the region above the ionosphere in which the magnetic field of the earth has a dominant control over the motions of gas and fast charged particles. This region is known to extend out to a distance of the order of 10 earth radii; it may appropriately be called the magnetosphere." This term is now used almost universally in referring to a large body of geophysical phenomena as well as to corresponding phenomena at other planets and other celestial objects (e.g., pulsars).

VIII.

The Argus Tests

As I have noted in the preceding chapter, another development of major importance to us was moving along in parallel with our work on the data from *Explorers I* and *III*. In December 1957 Nicholas Christofilos, a Greek engineer-scientist at the Livermore Radiation Laboratory, had proposed exploding one or more small nuclear fission bombs at a high altitude (~200 km) to test two effects that he envisioned: (*a*) the prompt enhancement of ionospheric ionization and the consequent disruption of radio communications at VHF frequencies and (*b*) the injection of large numbers of energetic electrons ($E_e \sim 2$ MeV) into durably trapped orbits in the earth's magnetic field. The electrons (and positrons) from the decay of radioactive fission products would be the principal source of both effects; an additional source of ionospheric influence would be the prompt ultraviolet, x- and γ-radiations from the burst itself.

In mid-April 1958 I informed Pickering and Panofsky of my by-then reasonably firm interpretation of the observations by *Explorers I* and *III*— namely that there was a huge population of electrically charged particles already present in trapped, Størmerian orbits in the earth's external magnetic field. In the context of our earlier studies of the primary auroral radiation, I considered it likely that these particles had a natural origin.

At this time, I was allowed to know the secret plans for the conduct of the high-altitude bomb tests, later called Argus. I also learned that, despite the absence of definitive information, some officials of the Atomic Energy Commission believed that the Soviet Union might have already conducted such tests. Panofsky suggested that, if this were in fact true, the radiation discovered by *Explorers I* and *III* might be of such artificial origin. A complementary line of thought was that, in either case, the Argus tests would provide the United States with the necessary competence to detect high-altitude nuclear bomb tests by the Soviet Union or by other countries.

Using unclassified information on the fission yield of nuclear (atomic) bombs and electron spectra of the radioactive decay products and estimates of injection efficiency and the geometry of geomagnetic trapping, I estimated the resulting intensities and spatial distribution of trapped electrons.

My principal interest was to have an opportunity to follow up on our earlier work by flying detectors of the proper dynamic range and of more discriminatory capability than that of the Geiger tubes on the original *Explorer*s. At the same time I was eager to participate in the Argus tests because of their apparent national importance and more particularly because of the possibility of distinguishing between a natural and an artificial population of geomagnetically trapped particles and of making direct observations of the residence times and diffusion rates of a known spectrum of electrons, injected at a known place at a known time. Ludwig, McIlwain, Ray, and I set to work designing, building, and testing a system of detectors and associated electronics for further satellite missions. Our principal working relationship in developing a practical payload for a Jupiter C vehicle was with Ernst Stuhlinger, Joseph Boehn, and Charles Lundquist of the Army Ballistic Missile Agency in Huntsville, Alabama, and with JPL.

Porter and Kaplan were instrumental in arranging for IGY sponsorship and U.S. Department of the Army support for our work as an extended part of the IGY satellite program. Such sponsorship put our work on an unclassified level, as was altogether proper, but it also shrouded the classified aspect. In short, IGY sponsorship was the truth, but not the whole truth.

On May 1, 1958, we received an informal go-ahead with assurance of financial support. The proposed tests were being planned for August and September, a considerable challenge to say the least.

JPL had upgraded the performance of the high-speed stages of the Jupiter C so that a significant increase in the mass of the payload was possible, but we continued to labor under very severe constraints of mass, power, and telemetry capacity. Our optimized design provided battery power for about two months of flight. Two satellite flights were planned, as were some nineteen rocket flights by the Kirtland Air Force Base and a variety of ground- and ship-based observing programs. Three rocket flights of nuclear bombs were planned in the Argus program. From a geomagnetic point of view the best site for the injection of electrons into durable orbits was near the geomagnetic equator in the South Atlantic. Because of the eccentricity of the earth's magnetic field, a site at that longitude could minimize the altitude at which injection had to occur, while an equatorial site could maximize the efficiency for injection in order to produce durably trapped orbits. Launching from a ship in an isolated site was desirable because it allowed the secrecy of the operation to be safeguarded. Two satellite launches and three bomb injections were judged to be the minimum effort to give reasonable assurance of success. The Navy's guided missile ship, the USS *Norton Sound*, which we had "initiated" with Aerobee rocket launchings in 1949, was selected to launch the rockets. Needless to say, intensive coordination of the efforts of hundreds of people was required to assure operational success. Herbert York, director of the Advanced Research Project Agency of the DOD, was the central person in this coordination. In addition to the three Argus bursts of small (~1 kiloton) bombs, the AEC planned tests of two large (~10 megaton) bombs at altitudes of 40–70 km over Johnston Atoll in the central Pacific.

Visitors to the University of Iowa during the spring and summer of 1958 were astonished to find that a crucial part of this massive undertaking had been entrusted to two graduate students and two part-time professors, working in a small, crowded basement laboratory of the 1909 Physics Building. But we knew our business and were in no way intimidated by representatives of huge federal agencies. We settled on an array of four basic radiation detectors: a thin plastic scintillator on the face of an end window photomultiplier, a thin slice of cesium iodide crystal on the face of another photomultiplier tube [McIlwain 1960a], and two miniature Geiger-Mueller tubes, one lightly shielded and the other heavily shielded (fig. 27). Every effort was made to provide the necessary dynamic range to cope with the intensity of the natural radiation and with the estimated intensity of the additional radiation that was to be artificially injected. The upgraded high-speed stages of the Jupiter C made it possible to plan an increase in inclination of the satellite orbit from the 33° of the orbits of *Explorers I* and *III* to 50° in order to provide improved coverage in latitude.

On July 1, 1958, Ludwig, McIlwain, and our new electronics engineer, Donald Enemark, took our completed No. 1 and No. 2 payloads to Hunts-

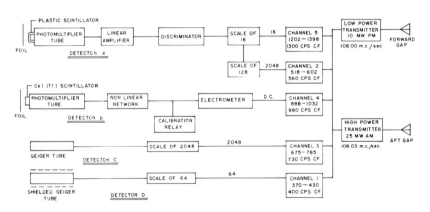

J. A. Van Allen, C. E. McIlwain, and G. H. Ludwig, "Satellite Observations of Electrons Artificially Injected into the Geomagnetic Field," *Journal of Geophysical Research*, vol. 64 (1959), p. 879, copyrighted by the American Geophysical Union.

Fig. 27. *Functional schematic of the four radiation detectors and associated circuitry of the payload of* Explorer IV *[Van Allen, McIlwain, and Ludwig 1959b]*.

ville for final environmental testing there and then to Patrick Air Force Base in Cocoa Beach, Florida, for spin balancing.

Explorer IV was launched successfully up the east coast of the United States on July 26, 1958. It carried Iowa payload No. 2, the best of four units that we had built. In my journal I wrote on July 29:

Apparatus seems to be working quite well! Confirms existence of 'soft radiation'—except it doesn't seem to be so soft (?).

And on July 30:

President Eisenhower yesterday signed the National Aeronautics and Space Act.

McIlwain brought our payload No. 3 (with a bad PM tube) back to Iowa City while Ludwig remained at Cape Canaveral checking out No. 4. The No. 1 payload had been fully checked out and was also available for flight on *Explorer V*. The entire system of orbit determination and data acquisition (all real time, no storage) had been improved markedly during the preceding six months, and by July 31 we had examined the data from several recorded passes by various stations. My journal summary was as follows:

General situation on data: Explorer IV (satellite 1958ϵ)
(1) All detectors working splendidly.
(2) 1.4 g cm^{-2} of Pb around the shielded counter reduces the counting rate only *mildly* at the higher altitudes, i.e., $\sim$ factor of 2!
(3) Detector A (pulse scintillation channel) giving fluxes $\sim 10^4$ (cm^2 sec sr)$^{-1}$ at ~ 1500 km altitude, latitude $\sim 12°$N (E $>$ 600 keV if electrons).
(4) Detector B (CsI detector) gives ~ 1 erg (cm^2 sec sr)$^{-1}$.
(5) Rapid altitude dependence between 600 $\rightarrow$ 1500 km observed as in Explorers I and III
(6) GM tube rates, shielded 184 sec^{-1}, unshielded 308 sec^{-1} at 1500 km, 12°S.

Based on data so far received and read (~ 6 to 8 passes here and there):
I. Previous results from [satellites 1958] α and γ well confirmed.
II. Evidence for a high energy tail or perhaps another phenomenon than aurora. However, Kinsey Anderson has previously had evidence for high energy ~ 500 keV electrons. Due for publication in Phys. Rev. on August 15.

On August 4 my wife and I packed our then four children into our station wagon and drove to Long Island, New York, for a long planned, but brief, vacation. The following is a verbatim transcript of my journal:

9 August 1958

I called McIlwain in Iowa City on Thursday evening with the main purpose of proposing that Detector B (The CsI crystal channel) be *heavily shielded* on Explorer V flight—i.e., that the 0.7 mg/cm² approx. of nickel foil be replaced by ~⅛" of aluminum.

Reasons as follows:

(a) Explorer IV seems to be providing quite satisfactory reading for the thin window case and there is little point to a straight repetition.

(b) The thin window case has no pertinence to the A.E.C. problem.

(c) The high intensity radiation may very well be dominantly *protons*. Injection mechanism may be—decay of albedo (upward moving from top of atmosphere) neutrons at high altitudes into *protons* and *electrons*. Since "lifetime" of trapped particles in the earth's magnetic field is probably limited dominantly by *scattering out of* trapped orbits which are mirrored at high altitudes to ones less inclined to the magnetic lines— which are therefore mirrored at lower altitudes and hence result in more rapid loss of energy by ionization—protons probably a *very* much greater lifetime than electrons of comparable energy. The factor may well be of the order of mass ratio—at least 2000 to one.

In order that a particle, trapped in a Størmer-Treiman lune, have a long life it should preferably be injected at high altitude where the atmospheric density is low. Cosmic ray processes best able to accomplish this are:

(a) Decay of μ-mesons to inject electrons of energies ~ or > 100 MeV.

(b) Decay of neutrons—a *thermal* neutron gives a β-ray spectrum of electrons with upper limit 782 keV *and* a very low energy proton (should calculate!). A *fast* neutron (whose decay in the vicinity of the earth is of course correspondingly reduced by its travel distance during its lifetime) gives a proton and an electron with energies in the laboratory system appropriate to the kinematics. Proton energy of the order of magn. of that of the neutron will result. (Should make detailed calculations.)

(c) Decay of neutrons from the sun in the vicinity of the earth has been suggested by J. Kasper as a possibility for injection. This is a conceivable possibility. But work of Bergstrahl and Perlow (?) suggests neutron flux from the sun negligible compared to that from the earth on basis of lack of day-night difference with BF_3 counters at 100,000 ft. (Should look at more critically! and perhaps devise a more discriminating experiment.)

(d) With thick shield over CsI detector we confine our observations to the more penetrating portion of the radiation.

(e) Explorer IV is working fine. Its data with the thin window over the CsI crystal will presumably have given good coverage of the physical situation under these conditions of observation. Hence V may as well and in fact more profitably should have a different arrangement.

———

We had previously agreed to reduce the *dimensions* of the plastic scintillator on Detector A. McIlwain advised that on the basis of rough but detailed calculations he had settled on dimensions as follows for the plastic scintillator:

$$\text{Thickness} \begin{cases} 0.070'' \\ \text{to} \\ 0.050'' \end{cases} \qquad \text{Diameter} \begin{cases} 0.070'' \\ 0.050'' \end{cases}$$

The scintillators which had already been cemented on the 6199s for payloads I and IV were machined down in our shop in situ without disturbing anything else.

The basic idea of this change was to reduce the detection efficiency for fast electrons (and fast protons) passing through the disc-shaped scintillator from the *side* (Explorer IV)—i.e. through the side walls, etc. But to retain ample sensitivity to lower energy electrons (A.E.C.) entering through the *aperture*!

Had also agreed to use a considerably thicker shield of Pb over the shielded G.M. tube to get a further test of hardness ($\sim 3/16''$ Pb).

The modified payload No. 4 was launched as *Explorer V* on August 24, 1958, but the final stage of the Jupiter C failed to ignite and an orbit was not achieved. Our apparatus functioned properly during its brief ten-minute flight before falling into the sea.

On the basis of the first few weeks of data from *Explorer IV*, we had advised ARPA of our discovery of a minimum in the previously present radiation when intensity was plotted against latitude. This finding was utilized in helping select the latitude for the Argus bursts so that the artificial radiation belts would enjoy the optimum prospect for detection. This choice of latitude turned out to be the best possible choice within the latitude range of *Explorer IV*, i.e., in the "slot" between the previously observed "inner" radiation belt and the newly discovered "outer" radiation belt.

The AEC/DOD task group successfully produced two bursts of 10-megaton yield bombs, called Teak and Orange, on August 1 and August 12 at approximate altitudes of 75 and 45 km, respectively, above Johnston Atoll in the Central Pacific. The three Argus bursts (about 1.4 kiloton yield) were produced successfully on August 27, August 30, and September 6 at altitudes of about 200, 250, and over 480 km at locations 38° S, 12° W; 50° S, 8° W; and 50° S, 10° W, respectively.

We observed with *Explorer IV* the effects of all five of the bursts in populating the geomagnetic field with energetic electrons. Despite the large yields of Teak and Orange, the incremental effects on the existing population of trapped particles were small and of only a few days lifetime because of the atmospheric absorption corresponding to the low altitudes of ignition.

The three higher-altitude Argus bursts produced clear and well-observed effects (fig. 28) and gave a great impetus to understanding geomagnetic

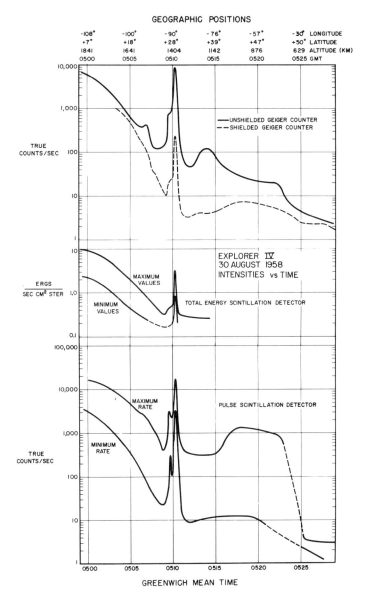

J. A. Van Allen, C. E. McIlwain, and G. H. Ludwig, "Satellite Observations of Electrons Artificially Injected into the Geomagnetic Field," *Journal of Geophysical Research*, vol. 64 (1959), p. 885, copyrighted by the American Geophysical Union.

Fig. 28. The narrow double spikes in the responses of the four radiation detectors were observed at about 0510 GMT on August 30, 1958, as Explorer IV *traversed the shells of energetic electrons injected into trapped orbits by the Argus I burst on August 27 and by the Argus II burst about three hours earlier on August 30 [Van Allen, McIlwain, and Ludwig 1959a].*

trapping. About 3% of the available electrons were injected into durably trapped orbits. The apparent mean lifetime of the first two of these artificial radiation belts was about three weeks and of the third, about a month. In all three cases a well-defined Størmerian shell of artificially injected electrons was produced. Worldwide study of these shells provided a result of basic importance—a full geometrical description of the locus of trapping by "labeled" particles. Also, we found that the physical nature of the Argus radiation, as characterized by our four *Explorer IV* detectors, was quite different than that of the pre-Argus radiation, thus dispelling the suspicion that the radiation observed by *Explorers I* and *III* had originated from Soviet nuclear bomb bursts.

During the approximate month of clear presence of the three artificial radiation belts, there was no discernible radial diffusion of the trapped electrons, thus permitting determination of an upper limit on the radial diffusion coefficient for such electrons. The gradual decay in intensity was approximately explicable in terms of pitch angle scattering in the tenuous atmosphere and consequent loss into the lower atmosphere.

A comprehensive ten-day workshop on interpretation of the Argus observations was conducted at Livermore in February 1959. The physical principles of geomagnetic trapping were greatly clarified at this workshop. To us, one of the principal puzzles had been the durable integrity of a thin radial shell of electrons despite the irregular nature of the real geomagnetic field and the existence of both radial and longitudinal drift forces resulting from gradients in the magnetic field intensity. We had previously understood the importance of the first adiabatic invariant of Alfvén in governing trapping along a given magnetic line of force and the effects of the radial component of the gradient of the magnetic field intensity B in causing longitudinal drift in an axially symmetric field. But the longitudinal component of the gradient of B seemed to imply irregular drift in radial distance and hence in radial spreading, contrary to observation. The puzzle was immediately solved by Northrop and Teller [1960; Northrop 1963; cf. Kellogg 1959b], who invoked the second and third adiabatic invariants of cyclic motion to account for the observations. These theorems had been proven previously by Rosenbluth and Longmire [1957] and applied to plasma confined by a laboratory magnetic field. A specific application of these principles was McIlwain's [1961] concept of the L-shell parameter for the reduction of three-dimensional particle distributions to two-dimensional ones—a concept that has permeated the entire subsequent literature of magnetospheric physics.

The adiabatic conservation and nonadiabatic violation of these three invariants have proved to be central to understanding trapped particle motion and to play a basic role in all of magnetospheric physics. In effect, they supplant the rigorous integral of motion found by Størmer for an axisymmetric magnetic field and make it possible to understand trapped particle motion and the diffusion of particles when the conditions for conservation of the three invariants are violated by time-varying magnetic and electric fields. The three invariants correspond to the three forms of cyclic motion, with quite different periods, into which the Størmerian motion of a charged particle in an approximate dipolar magnetic field can be analyzed. The first is the gyro motion of the particle around a field line; the second is the latitudinal oscillation of the guiding center (the center of the cylinder on which the helical motion of the particle occurs) of the particle's gyro motion; and the third is the time-averaged cyclic drift of the guiding center through 360° of longitude.

The Kirtland rocket measurements were generally consistent with our *Explorer IV* measurements but added important detail on particle identification and energy spectra. Also, atmospheric luminescence of auroral character was observed along the lines of force on which the bursts occurred; an artificial auroral display was observed at the northern geomagnetic conjugate point of the third burst; radar reflections from the auroral tubes of force were observed in all three cases; and a variety of transient ionospheric effects were detected. No electromagnetic (cyclotron) emission from the trapped electrons was observed by ground stations, a result consistent with estimates of the intensity relative to cosmic background.

The entire Argus operation was conducted in secret as were the reduction and interpretation of the observations, all under the general supervision of the Advanced Research Projects Agency of the Department of Defense. The secret level of security was maintained for at least six months even though the nature of the operation was known in a general way by perhaps a thousand persons (including the ship's company of the USS *Norton Sound*) and in detail by many participants and two *New York Times* reporters—Hanson W. Baldwin, journalist for military matters, and Walter Sullivan, science writer. In mid-March 1959 after much internal discussion, the Department of Defense and the Atomic Energy Commission decided to declassify the major features of the Argus tests. Detailed articles on the tests were written by Baldwin, Sullivan, and others and published in the *New York Times* on March 19, 20, and 26 and April 30, 1959.

The radiation detectors on *Sputnik III* (chapter 10) were operating during the period of the Teak high-altitude bomb test [Herz et al. 1960 cite data for July 19, 23, and 30 and August 4] and probably during the period including the Orange test [Vernov et al. 1959a]. But I have been unable to find any published record of Soviet satellite observations of the two artificially produced radiation belts resulting from these two tests.

J. G. Keys observed an auroral display in Apia (13° 48′ S, 171° 46′W) beginning at 10:51 UT of August 1, 1958, and a magnetic sudden commencement and reverse impulse ssc on the station magnetometer there beginning at 10:50 UT. Natural events of this character are exceedingly unusual at this station. Cullington [1958] attributed them to the Teak test over Johnston Atoll, noting the approximate magnetic conjugacy of Johnston Atoll and Apia. (The Teak and Orange tests had been reported by the

New York Times on August 2 and 13, 1958, respectively.)

Geomagnetic effects of the Teak and Orange tests were also reported or discussed theoretically or both in early papers by Fowler and Waddington [1958]; Maeda [1959]; Kellogg, Ney, and Winckler [1959]; Malville [1959]; Lowrie, Gerard, and Gill [1959]; Mason and Vitousek [1959]; McNish [1959]; and Matsushita [1959]. The fireball and resulting aurorae from Teak were photographed comprehensively from Mt. Haleakala, Maui, Hawaii, by W. Lang [Steiger and Matsushita, 1960]. Mason and Vitousek [1959] also reported magnetic effects of a British nuclear bomb burst near Christmas Island on April 28, 1958, at an altitude of less than or about 18 km. Geomagnetic, auroral, and other atmospheric effects of the three Argus bursts were observed by Newman [1959], Peterson [1959], and Berthold, Harris, and Hope [1960], all of whom were participants in the test program; but apparently no one else identified such effects, except retrospectively after the March 1959 public release of information on the tests. For example, the Soviet geomagnetician V. A. Troitskaya [1960] found anomalous geomagnetic effects in telluric current records at the times shown in the fourth column of table 1.

TABLE 1

Date 1958	Burst	Actual time of Burst	Times of Magnetic Disturbance —Troitskaya
August 1	Teak	10:50:05 UT	— UT
August 12	Orange	10:30:08	10:30:08
August 27	Argus I	02:27:52.6	02:27–02:27:29
August 30	Argus II	03:17:33.8	03:17:34
September 6	Argus III	22:12:33.4	22:12:37

NOTE: The times listed in the third column are from the official U. S. records of the tests. The actual times were rounded to the nearest ten minutes in the data available to Troitskaya when she did her search.

The National Academy of Sciences agreed to sponsor an open symposium emphasizing the scientific aspects of the tests and the relevance of the results to understanding the dynamics of geomagnetic trapping and related matters. This "Symposium on Scientific Effects of Artificially Introduced Radiations at High Altitudes" was held on April 29, 1959. The seven papers presented at that time were:

"Introductory Remarks"—Richard W. Porter

"The Argus Experiment"—N. C. Christofilos

"Satellite Observations of Electrons Artificially Injected into the Geomagnetic Field"—James A. Van Allen, Carl E. McIlwain, and George H. Ludwig

"Project Jason Measurements of Trapped Electrons from a Nuclear Device by Sounding Rockets"—Lew Allen, Jr., James L. Beavers, II, William A. Whitaker, Jasper A. Welch, Jr., and Roddy B. Walton

"Theory of Geomagnetically Trapped Electrons from an Artificial Source"—Jasper A. Welch, Jr., and William A. Whitaker

"Optical, Electromagnetic, and Satellite Observations of High-Altitude Nuclear Detonations, Part I"—Philip Newman

"Optical, Electromagnetic, and Satellite Observations of High-Altitude Nuclear Detonations, Part II"—Allen M. Peterson

All of these papers were subsequently published in the August 1959 issues of the *Journal of Geophysical Research* and the *Proceedings of the National Academy of Sciences* (see individual references in the bibliography).

The combination of all this initial work and of many subsequently published papers and still-classified reports clarified many aspects of magnetospheric physics and gave a great impetus to the subject. It also showed conclusively that the trapped radiation observed by *Explorers I, III*, and *IV* before the tests was different in character than was the radiation produced by nuclear bursts and was therefore of natural origin.

In early July 1959 I attended the Cosmic Ray Conference of the International Union of Pure and Applied Physics in Moscow and there gave a summary lecture (subsequently published in Russian) on the geomagnetically trapped corpuscular radiation. The entire subject, including the results of the widely publicized Argus tests, was received with keen interest. I was given a tour of Soviet laboratories and shown, among other things, the prototype instruments for *Sputniks II* and *III*. Leonid I. Sedov, chairman of the Interdepartmental Commission on Interplanetary Communications of the Astronomical Council of the USSR Academy of Sciences, invited me to give a seminar at the USSR Academy of Sciences. Being willing but a bit apprehensive, I asked two of the other U.S. delegates to the conference, John A. Simpson and George W. Clark, to accompany me. We were driven by car from Moscow University to the rather remotely located academy building. There, I spoke and showed slides for over an hour, with pauses for translation by the physicist G. I. Galperin, to a selected and very expert audience. Following this, I responded to a host of searching and critical questions on all aspects of the work, including the effects of high-altitude nuclear bomb bursts and the evidence for our claim that the pre-Argus radiation was of natural origin. It became clear that many Soviet scientists had had the reciprocal suspicion that the latter radiation had been injected into the geomagnetic field by clandestine U.S. bomb bursts at high altitude, either in late 1957 or early 1958.

An interesting sequel of this occasion was that in the autumn of 1959, Sedov, A. A. Blagonravov, V. I. Krassovskii, and a young Russian translator visited our laboratory at the University of Iowa on my invitation. They were, I think, especially impressed by the miniaturization and low power requirements of our instrumentation, which was in striking contrast to that flown on early Soviet spacecraft. Sedov gave a general university lecture on their plans for lunar flights, and Krassovskii gave a physics department colloquium talk on his observations of low-energy particles with *Sputnik III*.

At a dinner for the four visitors, which my wife had arranged at our home, Blagonravov entertained our young children by letting them listen to a ticking of his big pocket watch, a genial grandfatherly activity that I found to be an interesting footnote to his reputation as a tough-minded lieutenant general of the Soviet Army, who had supervised production of small arms in the USSR during World War II. At breakfast in a restaurant the following morning I encountered Blagonravov reading the university student newspaper, the *Daily Iowan*, to his colleagues.

IX.

Early Confirmations of the Inner Radiation Belt and Discovery of the Outer Radiation Belt—Explorer IV *and* Pioneers I, II, III, IV, *and* V

As described in the previous chapters, *Explorer IV* was the first satellite to carry a set of radiation detectors that had been designed with knowledge of the presence of the geomagnetically trapped radiation and of its intensity [Van Allen, McIlwain, and Ludwig 1959b]. For example, the Geiger-Mueller tubes (Anton Type 302) that we selected for *Explorer IV* had an effective cross-sectional area one hundred times smaller than those on *Explorers I* and *III*. The two scintillation detectors had correspondingly small geometric factors and were directional in order to study the angular distribution of the radiation [McIlwain 1960a]. All detectors had the greatest feasible dynamic ranges. Also, the orbit of *Explorer IV* was inclined to the geographic equator by 50°, thus providing latitude coverage that was much broader than that provided by the 33° inclination orbits of *Explorers I* and *III*. A worldwide network of twenty-two regular telemetry stations received data in real time.

Explorer IV yielded a massive body of data on the natural trapped radiation as well as on that injected by the five nuclear bomb bursts—Teak, Orange, and Argus I, II, and III—during the period from launch on July 26 to September 19, 1958. All detectors operated properly and remained within their dynamic ranges. Data were obtained for five days before the Teak test on August 1 and for thirty-one days before the first Argus test on August 27. The effects of all five of the bursts were clearly recognized in the data and were easily segregated from the effects of the natural radiation.

In December 1958 the Iowa group completed a paper on *Explorer IV* observations of the natural radiation, omitting those of the artificially introduced radiation. This paper, which was published in the March 1959 *Journal of Geophysical Research* [Van Allen, McIlwain, and Ludwig 1959b], gave the first comprehensive body of observations on the subject and provided massive confirmation of the results of *Explorers I* and *III*. The initial range of altitude was from 262 to 2,210 km. The diverse detectors on *Explorer IV* yielded both particle intensities and energy fluxes as well as angular distributions and crude intensity-range data for the radiation, all as a function of position along the orbit. Among other findings were the facts that at constant altitude there was a relative minimum value of intensity (the "slot") at about 48° latitude, both north and south, and an apparent high-latitude boundary of the trapped radiation at between 65° and 70° latitude. The radiation at latitudes less than 48° was markedly more penetrating than that at higher latitudes. These two findings supported our later *Pioneer III* findings of two major radiation belts having particle populations of distinctively different character. The possibility of two distinct radiation belts was considered at the time that we prepared the *Explorer IV* paper but was omitted in the final draft in favor of a single belt with a low-altitude slot at 48°. This mistake in our conjectural extension of isointensity contours was made clear by *Pioneer III* data obtained only a few days after submission of the *Explorer IV* paper. The mistake was not corrected in proof in order to preserve the integrity of the original paper.

Despite the variety of information obtained with the four different detectors on *Explorer IV* including temporal variations [Rothwell and McIlwain 1960], it was frustratingly difficult to reach firm conclusions on the identity and energy spectra of the particles responsible for the responses of the detectors. Also, we fully realized that much of the trapped radiation might lie at energies too low to be registered by our detectors. Because of the universal presence of electrons in ionized matter and because of the dominant abundance of hydrogen in astrophysical matter, we made the plausible assumption that the trapped radiation consisted of an admixture of electrons and protons having unknown energy spectra that might very well be different from each other and that were clearly a function of position in space. We made many different trial assumptions but were unable at that time to demonstrate any clear conclusions on these matters. We did state sample absolute intensities based on three alternative interpretations: (*a*) penetrating protons, (*b*) penetrating electrons, and (*c*) nonpenetrating electrons (via bremsstrahlung); and we suggested that the radiation might well be a mixture of all three, with the mix a function of position.

In a memorandum of May 23, 1958, entitled "Radiation Measurements with Lunar Probes," I outlined the desirability of observing the full radial structure and outer boundary of the radiation belt by means of an approximately radial scan that would be provided by a trajectory leading to the moon. This proposal was viewed favorably by the U.S. National Committee for the International Geophysical Year, and we again joined with the Jet Propulsion Laboratory and the Army Ballistic Missile Agency in preparing suitable payloads for the planned flights on Juno II launch vehicles (upgraded versions of Jupiter C).

We were also invited to help prepare radiation detectors for two Air Force moon flights, using Thor-Able launch vehicles. McIlwain equipped a small Anton type 706 ionization chamber with a d.c. logarithmic amplifier for the scientific payload of each of the two spacecraft *Pioneers I* and *II*. In order to provide against any possibility of saturation, he designed the system for a dynamic range of 1 to 10^6 roentgen per hour. *Pioneer I* was launched on October 11, 1958, and reached a maximum geocentric distance of about 19 earth radii before falling back to the earth, thus making the first passage through the radiation belt region to altitudes above 2,200 km. Despite some instrumental difficulties, Rosen et al. [1959] reported:

Analysis of the observed data indicates that the following deductions may be made:

1. The level of ionizing radiation rises to a peak and then drops to a very low level as the altitude increases. This represents the first experimental verification of the existence of a confined radiation zone of the type postulated by Van Allen and others [1958].

2. In the altitudes ranging from approximately 4,000 to 24,000 km the level of ionizing radiation is in excess of 2 roentgens/hour. This result yields a quantitative measure of the depth of the radiation belt in the region of 20°N.

3. In the vicinity of 20°N the peak of the radiation belt occurs at an altitude of 10,000 ± 2,000 km.

4. The maximum radiation level in this region is 10 roentgens/hour.

5. Although experimental errors were relatively large, it appears that a fine structure exists in this region.

6. The quantitative results given above are based upon the stated assumption that the chamber pressure was approximately 1.58 atmospheres during the first four hours of flight. It is also worthwhile to consider the results of an increased leak rate such that the pressure in the chamber at launching was atmospheric. In this case, the calculated sensitivity would be decreased by another factor of 1.6, and the radiation levels recorded in Figure 2 [of the quoted paper] would have to be increased by a factor of 1.6.

The third stage of the vehicle for *Pioneer II* failed to ignite following an otherwise successful launch on November 7, 1958, from Cape Canaveral. Nonetheless, the payload including its radiation detecting ionization chamber was propelled to a maximum altitude of 1,550 km and operated properly. In approximate agreement with *Explorer IV*'s data, the radiation intensity at a nearly constant altitude of 1,525 km increased by a factor of

13 as the spacecraft moved southward from 31° to 24°N. More important, the combination of absolute particle intensity data from the Geiger tubes on *Explorer IV* and data from the ionization chamber on *Pioneer II* yielded a rough determination that the average specific ionization was about five times its minimum value for a charged particle. In consideration of this finding and the necessary range of the radiation, it was concluded that the responsible radiation must be dominantly protons of energy ~120 MeV [Rosen, Coleman, and Sonett 1959]. A similar conclusion on somewhat different grounds was reached by Simpson, who had a wide-angle triple-coincidence array of semiproportional counters shielded by 5 g cm^{-2} of lead on *Pioneer II*.

Another Iowa student, Louis A. Frank, and I prepared a pair of miniature Geiger-Mueller tubes and associated power supplies and electronic circuitry with large dynamic range for flight on the first two ABMA moon flights. One of the tubes was the primary radiation detector and the other, smaller one was arranged as an "ambiguity-resolver" in case the radiation was sufficiently intense to drive the first tube over the top of its characteristic curve of apparent counting rate versus true counting rate. The telemetry transmitter (960.05 MHz) and antenna, battery pack, and payload structure and shell were built by the JPL [Josias 1959], which also conducted the necessary environmental tests (acceleration, vibration, and thermal vacuum) and established two special telemetry receiving stations—one near Mayaguez, Puerto Rico (10-ft. dish antenna), and the other at Goldstone Lake, California (86-ft. dish). *Pioneer III* was launched from Cape Canaveral on December 6, 1958. Earth escape velocity was not achieved. The payload reached a maximum geocentric distance of 107,400 km (17 earth radii) and fell back to the earth on the following day. Excellent data were obtained on both the outbound and inbound legs of the trajectory. From our point of view the flight was much more valuable than it would have been if it had flown to the moon because of the two quite different paths through the radiation belt that occurred. This flight established the large-scale features of the distribution of geomagnetically trapped particles and, in conjunction with the lower altitude data from similar detectors on *Explorer IV*, clearly defined two major, distinct radiation belts, both of toroidal form encircling the earth with their planes of symmetry at the geomagnetic equatorial plane [Van Allen and Frank 1959a; Van Allen 1959c]. The inner belt exhibited maximum intensity (as measured with a GM tube shielded by 1 g cm^{-2} of material) at a radial distance of 1.4 earth radii, and the outer, more intense and much larger belt exhibited a maximum intensity at about 3.5 earth radii. A "slot" or local minimum intensity was observed between the two major belts. A meridian cross-section of isointensity contours was prepared using *Pioneer III* and *Explorer IV* data (figs. 29, 30). This diagram has continued to have essential validity throughout all subsequent work.

A second payload of similar equipment was launched as *Pioneer IV* on

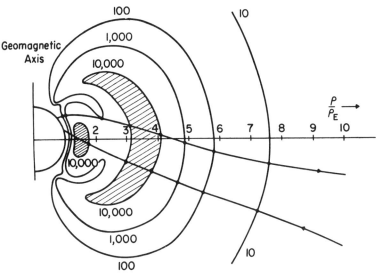

J. A. Van Allen, "The Geomagnetically-Trapped Corpuscular Radiation," *Journal of Geophysical Research*, vol. 64 (1959), p. 1684, copyrighted by the American Geophysical Union.

Fig. 29. A meridian cross-section of contours of equal intensity of geomagnetically trapped radiation based on data from Explorers I, III, *and* IV *and* Pioneer III. *The semicircle at the left represents the earth, and the two undulating curves that traverse the diagram represent the outbound (upper curve) and inbound (lower curve) trajectories of* Pioneer III. *The labels on the contours are counts per second of a lightly shielded miniature Geiger-Mueller tube. The linear scale of the diagram is in units of the earth's equatorial radius (6,378 km). The two distinct regions of high intensity (cross-hatched) are the inner and outer radiation belts, separated by a region of lesser intensity called the slot [Van Allen and Frank 1959a].*

Courtesy of J. A. Van Allen.

Fig. 30. An artist's three-dimensional conception of the earth and the inner and outer radiation belts as derived from figure 29.

March 3, 1979 (fig. 31). Earth escape velocity was achieved, but the payload missed the moon by a radial distance of 62,000 km and continued into a heliocentric orbit. Data were received by the Cape Canaveral, Mayaguez, and Goldstone stations. A valuable addition to telemetry reception for this flight was provided by the 250-ft. Jodrell Bank dish, which obtained data of good quality to a geocentric distance of 658,300 km (103 earth radii), far beyond the moon's orbit. The *Pioneer IV* radiation data confirmed the inner zone/outer zone structure of the geomagnetic trapping region and showed a much greater intensity in the outer zone than that in early December 1958 [Van Allen and Frank 1959b; Snyder 1959]. This enhanced intensity was attributed to unusually strong solar activity during the few days preceding the flight of *Pioneer IV*, whereas there was an especially quiet period before and during *Pioneer III*'s flight. Also, the

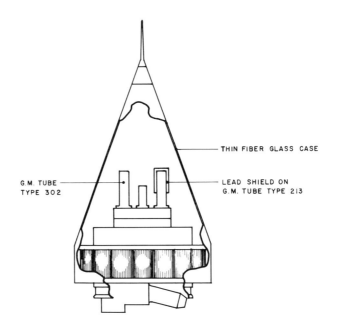

Reprinted by permission from *Nature*, vol. 184, no. 4682, p. 219. Copyright © 1959 Macmillan Journals Limited.

Fig. 31. Physical arrangement of radiation detectors in the payload of Pioneer IV *(nearly identical to that of* Pioneer III*). The total mass of the payload was 6.1 kg, and the outside diameter of the lower (cylindrical) portion of the shell was 23 cm [Van Allen and Frank 1959b].*

interplanetary value of the cosmic-ray intensity was well determined at points remote from the earth for the first time. No effect of the moon was detected.*

The last of the early series of Pioneer spacecraft, *Pioneer V*, was launched on an earth-escape trajectory on March 11, 1960. It contributed information on the earth's distant magnetic field, and it yielded several weeks of observations of the interplanetary magnetic field [Coleman, Davis, and Sonnett 1960; Coleman, Sonnett, and Judge 1960] and of energetic solar particle events [Arnoldy, Hoffman, and Winckler 1960a; Fan, Meyer, and Simpson 1960a]. The high-intensity particle events of late March and early April 1960 were observed simultaneously at high latitudes by the low-altitude earth orbiter *Explorer VII*, as was the associated, marked reduction in cosmic-ray intensity (a Forbush decrease) [Van Allen and Lin 1960]. A major magnetic storm was observed by ground-based magnetometers [Chinburg 1960], and *Explorer VII* also observed a major decrease in outer-zone particle intensities. This decrease was followed by recovery to much enhanced values on a time scale of a few days. The entire bundle of observations constituted one of the most convincing early examples of the control of the particle population of the outer radiation belt by solar corpuscular streams. It also added direct evidence to the prevailing hypothesis [Morrison 1956] that fluctuations of cosmic-ray intensity, as observed on the surface of the earth, are caused by solar corpuscular streams in interplanetary space [Fan, Meyer, and Simpson 1960c].

During 1958 the Iowa satellite research group consisted principally of three very talented and assiduous graduate students (Ludwig, McIlwain, and Kasper), an able, tough-minded young undergraduate student (Frank), a recent Ph.D. assistant professor (Ray), and myself. Both Ray and I were simultaneously teaching courses, while I was also serving as head of the Department of Physics and as a member of various national planning committees and panels. We had the services of three skilled instrument makers, J. George Sentinella, Edmund Freund, and Robert Markee. Within a fourteen-month period we provided the principal scientific instrumentation for *Explorers I, II, III, IV*, and *V* and for *Pioneers I, II, III*, and *IV*. Of these nine missions, seven yielded valuable radiation data; only *Explorers II* and *V* failed to do so—both because of failure of launch vehicles. The work of preparing the equipment for *Explorers I, II*, and *III* began in 1956 and was carried forward principally by Ludwig, who devised many novel circuits using the then new technology of transistor electronics and designed and nurtured the development of the miniature magnetic tape recorder. He temporarily transferred to the Jet Propulsion Laboratory from November 1957 to April 1958 to adapt the Iowa apparatus to the payloads

---

*The paper in *Nature* on our *Pioneer IV* results carried a publication date (July 25, 1959) that preceded the date of submission. This unusual circumstance resulted from a printer's strike and the journal's subsequent effort to fill in delayed issues.

of *Explorers I*, *II*, and *III* (Deal I, Deal II A, and Deal II B). Apparatus for the other five satellites and moon shots was conceived, designed, built, tested, and calibrated by Ludwig, McIlwain, Frank, and me after late April 1958. Ray and Kasper provided a wealth of theoretical help and assistance with data reduction, orbital calculations, and the like. (Retrospectively, I consider 1958 the most intense and rewarding year of my professional life.) Our work was supported in part by the state of Iowa and in part by the Army Ordnance Department, the Office of Naval Research, the Jet Propulsion Laboratory, and the National Science Foundation. But despite the multiplicity of sources of support, we had a minimal burden of paper work and enjoyed extraordinarily free and entrepreneurial working circumstances. It is scarcely necessary to remark on the contrast with the 1983 circumstances of conducting space research, which require massive proposing, defending, documenting, reproposing, reporting, financial- and activity-accounting, and typically a time lapse of a decade between the start of a project and its consummation. A significant part of such a long time lapse is attributable to technical sophistication and real developmental work, but perhaps one-half of it is of an administrative and, hence, avoidable nature. There must be a way to cut through the administrative thicket, but those of us who are active in space research at the present date have not been able to find it, even though we can envision an immense increase in professional productivity if we could.

X.

Related Work with Sputniks II *and* III *and* Luniks I, II, *and* III

After World War II the Soviet Union began developing high-performance rockets in a manner roughly parallel to, but largely independent of, such developments in the United States. Also, Soviet scientists adopted rockets for upper atmospheric and high-altitude research, though most U.S. workers between 1946 and 1955 were only dimly aware of these efforts [Vernov et al. 1958a]. The International Geophysical Year opened up lines of professional communication and revealed the relatively advanced plans that the Soviet Union had for launching artificial satellites for scientific purposes [Krieger 1958]. Those who had not taken these plans seriously were rudely surprised by the successful launching of *Sputnik I* on October 4, 1957, early in the IGY. This satellite carried no on-board scientific instruments, but the high power of its radio transmitter and the sagacious choice of frequencies (20.005 and 40.002 MHz) immediately impressed scores of radio professionals and amateurs and, through them, many millions of other persons. In addition, study of the ionospheric refraction and absorption of the radio signals provided a wealth of ionospheric information, while the precise determination of the satellite's orbital elements and their rates of change provided a great stimulus to the field of celestial mechanics, which had been rather quiet for several decades.

Sputnik II, launched on November 3, 1957, carried a pair of simple cosmic-ray intensity instruments in each of which the detecting element was a charged-particle counter (Geiger tube) 10 cm in effective length and 1.8 cm in diameter, shielded by about 10 g cm^{-2} of unspecified material. Radiation data were obtained for seven days. The orbit had perigee and apogee altitudes of 225 and 1670 km, respectively, and an inclination of 65° to the earth's equator. According to Vernov et al. [1958b], the purpose of their investigation was to obtain data on the distribution of cosmic-ray intensity over the whole of the earth, an objective paralleling that of the Iowa group in its 1955 proposal for early U.S. satellite flights. Vernov et

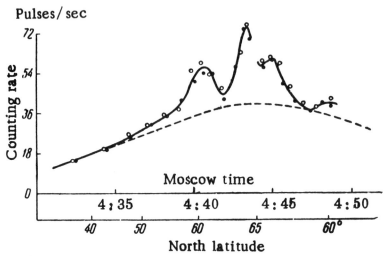

From S. N. Vernov and A. E. Chudakov, "Investigations of Cosmic Radiation and of the Terrestrial Corpuscular Radiation by Means of Rockets and Satellites," *Soviet Physics USPEKHI*, vol. 3 (1960), p. 232, by permission of the American Institute of Physics.

Fig. 32. The counting rates of two Geiger-Mueller tubes on Sputnik II *as a function of time (and latitude) at an altitude of about 300 km, showing the noteworthy cosmic-ray burst on November 7, 1957. The dashed curve gives average counting rates observed on other traversals of the same region.*

al. published a preliminary account of their observations in mid-1958, but I have not been able to find any subsequent accounts by them containing more observations. Measurements over the Soviet Union in the altitude range 225 to 700 km, the latitude range 40° to 65°N, and the longitude range 25° to 143°E were reported. Reliable data were obtained on the latitude and altitude dependence of cosmic-ray intensity. On a particular pass on November 7 a brief "burst" of substantially greater (~50%) than average intensity was observed coherently by both detectors at latitudes greater than 58°N (fig. 32). This effect was termed a "cosmic-ray burst," but no physical interpretation was suggested. It is clear that the instruments of Vernov et al. on *Sputnik II* could have revealed the presence of the radiation belts of the earth if data had been available from lower latitudes and high altitudes, despite the heavy shielding of the detectors. It may be noted that the magnetic center of the earth is displaced 450 km from its geometrical center toward a longitude of about 150°E. Hence, over central Russia the lower boundary of the radiation belt is at a much greater (~600 km) altitude than it is over, say, South America. Additionally, the center of the auroral zone lies at higher latitudes over Russia than the maximum latitude of *Sputnik II*'s orbit. The observed "cosmic-ray burst" had a structure corresponding to a time scale of the order of a minute or to a spatial scale of the order of a few hundred km. With the benefit of later knowledge, it appears very unlikely that this effect can be attributed to particles reaching the earth from interplanetary space, whether of solar or galactic origin. But it is plausibly identified with the fluctuating low-altitude horn of the outer radiation belt, an effect observed repeatedly by Lin and Van Allen with *Explorer VII* in 1959–60 [Van Allen 1962]. In this case, the response of the detectors would be to bremsstrahlung generated by bombardment of the outer skin of the satellite by electrons.

There is no evidence, so far as I can find, that Soviet scientists were aware of the existence of the earth's radiation belts before my May 1, 1958, public report of *Explorer I / III* observations. *Sputnik III* was launched on May 15, 1958, into an orbit very similar to that of *Sputnik I*. An extensive article on the spacecraft, its orbit, and the scientific purposes of the mission was published in *Pravda* on May 18, 1958. There was no mention of geomagnetically trapped radiation as a subject for investigation. Vernov et al. [1959a] obtained a valuable body of radiation data using a 4.0 × 3.9 cm NaI crystal mounted on a photomultiplier tube whose counting rate (~35 keV threshold) as well as its anode and intermediate dynode currents were measured. At latitudes of 55° to 60°N over the Soviet Union, they found variable but high intensities of radiation (usually exceeding the dynamic range of their detector). They stated:

From a comparison of this ionization increase with the increase of the counting rate, it is possible to estimate the energy of the bremsstrahlung photons and, consequently, the energy of the electrons. . . . the most probable value for the energy of the electrons responsible for this effect is about 100 kev . . . we estimate that the electron flux is 10^3–10^4 particles/ cm^2 sec sterad.

The authors conclude this section of their paper as follows:

At present, it is difficult to give a complete interpretation of the observed electron component. It is possible that these electrons are accelerated near the earth by electric fields analogous to those that are assumed to exist in aurorae. However, it is also possible that the electron component originates away from the earth, for example on the sun, and despite the low energy of the particles, penetrates through the terrestrial magnetic field, if this field departs from that of an ideal dipole.

On a purely observational basis the *Sputnik III* data represented discovery of the earth's *outer* radiation belt inasmuch as they were acquired before those of *Explorer IV* and *Pioneer III*. However, the interpretation of Vernov et al. did not encompass the idea of geomagnetic trapping for what they called the "electron component of the cosmic rays in the polar region."

Data from *Sputnik III* were also recorded by classified stations in the United States and by groups in Australia [Herz et al. 1960] and in Alaska [Basler, Dewitt, and Reid 1960]. The former observed four southbound passes over eastern Australia in the period between July 19 and August 4, 1958; it was preoccupied with diagnosis of the operation of the equipment, particularly the scintillation crystal and its associated photomultiplier and data-formating system, but reported apparently reliable results on the strong latitude dependence of intensity as the satellite passed from the slot region into the outer belt. It also reported marked temporal variations of intensity in this region. The Alaskan group recorded data from the sodium iodide crystal for sixty-two passes from "18 May until the partial failure of the telemetering system [presumably on the satellite] on 17 June 1958." It found a relatively coherent increase of total radiation intensity with altitude in the range 250 to 540 km but were considerably plagued by the time-dependent afterglow of the crystal, which was attributed to recent passage through the inner radiation belt.

The next section of the important paper of Vernov et al. [1959a]—first reported in preliminary form at the fifth meeting of the CSAGI, held from July 31 to August 9, 1958—is entitled "The High-Intensity Zone in the Equatorial Regions." Therein, the authors confirm our *Explorer I / III* results and add significant detail on latitude and altitude dependence of intensity (fig. 33). They also found the radiation intensity in the equatorial region to have approximate north-south symmetry relative to the geomagnetic equator and to be stable in time. The time stability at low latitudes was contrasted with the marked temporal variability at high latitudes. They estimated that the energy deposition rate in the NaI crystal was of the order of 5 ergs $(cm^2\ sec)^{-1}$ at an altitude of 1,200 km near the equator. Also they remarked:

Whatever the mechanism for the production of particles in the equatorial region, it is obvious that storage of particles is the most important factor

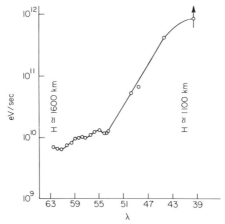

Reprinted with permission from *Planetory and Space Science*, vol. 1, S. N. Vernov, A. E. Chudakov, E. V. Gorchakov, J. L. Logachev, and P. V. Vakulov, "Study of the Cosmic-Ray Soft Component by the 3rd Soviet Earth Satellite," Copyright 1959, Pergamon Press, Ltd.

Fig. 33. A sample curve of the dependence of radiation intensity on geomagnetic latitude λ, *observed as* Sputnik III *passed into the inner radiation belt. H is the approximate altitude of the satellite [Vernov et al. 1959a].*

contributing to this effect. Evidence for this is provided by the concentration of particles in the equatorial zone where, at a sufficiently high altitude, they can move for a very long time.

At the end of this section of the paper, the authors conclude:

It appears that in this equatorial zone conditions are ideal for the oscillation of particles in the magnetic field of the earth and loss of particles is possibly determined only by the ionization and radiation losses. In this case, the particles have a very long lifetime (of the order of a year), so that even a weak injection mechanism, such as the decay of neutrons ejected from the atmosphere through the action of cosmic rays, will be found adequate for the explanation of the observed intensity.

This conclusion clearly embraces the concept of geomagnetic trapping as appropriate for the equatorial region and suggests that the decay of neutrons produced in the atmosphere by cosmic rays provides a possible source. However, no observational identification of the nature of the particles was achieved nor was any information on intensity versus range obtained. Also, it was not clear whether the authors were thinking of low-energy or high-energy neutrons or of electrons or protons or both as significant decay products. A much more comprehensive paper on results of *Sputniks II* and *III* was published later by Vernov and Chudakov [1960], but no new *Sputnik II* data were presented.

On *Sputnik III* there were two other energetic particle detectors, thin ZnS(Ag) fluorescent screens 2×10^{-3} g cm^{-2} in thickness covered with aluminum foils 8×10^{-4} g cm^{-2} and 4×10^{-4} g cm^{-2} in thickness, respectively, prepared by V. I. Krassovskii [1960] and Krassovskii et al. [1960, 1961]. With these, they discovered the presence of electrons of energy about 10 keV at altitudes of $\sim 1,900$ km over the southern part of the Pacific Ocean. Directional energy fluxes of such electrons were found to be of the order of 100 ergs (cm^2 sec)$^{-1}$ and greater, and the "majority of the recorded electrons move near the plane perpendicular to the magnetic line of force. . . . the recorded currents were explained by an oscillatory motion of the electrons along the magnetic line of force." A net energy flux of the order of 1 erg (cm^2 sec)$^{-1}$ into the F-layer of the atmosphere was estimated from the data. Marked variations in time or position were found.

An important new body of observations of the density of ions in the upper ionosphere (500 to 1,000 km) was obtained by Gringauz, Bezrukikh, and Ozerov [1961] using "ion traps" on *Sputnik III*. The developmental work for this instrument was the basis for the design of many future "ion-trap" instruments as well. *Sputnik III* also carried a system of magnetometers by Dolginov, Zhuzgov, and Pushkov [1960] that were specially developed for space flight. They operated successfully but provided no basically new geomagnetic field information, though they were useful for determining the angle between the magnetic vector and the axes of particle detectors.

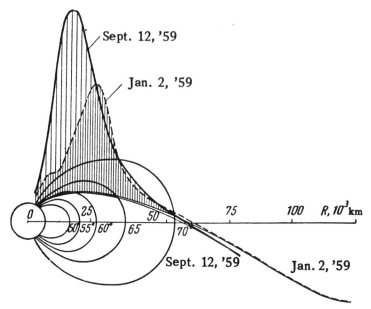

Reprinted from *Doklady Akademii Nauk SSR*, vol. 130 (1960), by permission of the American Institute of Physics.

Fig. 34. A diagram from Vernov et al. [1961] showing the trajectories of the first and second Soviet cosmic rockets (Luniks I and II) and, by the overlaid plots, the radiation intensity observed along these trajectories. R is the radial distance from the center of the earth. The numbers below the horizontal axis label the latitudes of the feet of the magnetic lines of force shown.

The next body of Soviet data relevant to the magnetosphere came from three "cosmic rocket" or moon flights (also called *Luniks I, II* and *III*) launched on January 2, September 12, and October 4, 1959, respectively. The first of these missed the moon by a radial distance of 7,700 km; the second crashed into the moon, becoming the first man-made object to do so; and the third flew around the moon (photographing the far side for the first time) and continued in a loose earth orbit until reentry on April 20, 1960. On the basis of published accounts it appears that *Lunik II* yielded the most important particles and fields data [Vernov et al. 1959b, 1961; Vernov and Chudakov 1960] (fig. 34).

The radiation detectors of Vernov et al. on *Lunik II* were six Geiger tubes and three scintillation counters in a variety of physical arrangements and with various electronic thresholds. In broad terms, the observed structure of the outer radiation belt confirmed what the American equipment on *Pioneers III* and *IV* had revealed previously (fig. 35). The diversity of detectors added valuable, though not altogether unambiguous, information. Combining results from *Luniks I* and *II*, Vernov et al. favored the interpretation that at the maximum of the outer belt there were:

(a) virtually no particles having a range greater than several grams per cm² (i.e., electrons of energy greater than 5 MeV or protons of energy greater than 30 MeV);

(b) an intensity of $\sim 10^{10}$ (cm² sec)$^{-1}$ of electrons in the energy range 20–50 keV.

(c) an intensity of $\sim 5 \times 10^5$ (cm² sec)$^{-1}$ of electrons of ~ 2 MeV energy (or protons of energy ~ 10 MeV); and

(d) a comparable intensity of electrons in the 0.1 to 1.0 MeV energy range.

The outer boundary of the earth's radiation belt was observed at a radial distance of about 70,000 km. No significant increase in the interplanetary intensity level was observed as the payload approached the moon, the last observation being obtained at an altitude of 1000 km above the moon's surface.

Kurnosova et al. [1961] had a separate radiation instrument containing two Cerenkov detectors on *Lunik II*. The principal purpose of this instrument was to search for α-particles and heavier ions. The results of this search within the earth's magnetosphere were inconclusive, but good results on the primary cosmic radiation were obtained enroute to the moon after passing out of the earth's influence. An auxiliary feature of the detectors was a "radiation indicator" utilizing the sensitivity of the photomultiplier tube itself to bremsstrahlung of energy greater than 15–20 keV as well as to electrons of energy greater than 2 MeV. This survey meter yielded a maximum intensity at about 17,000 km radial distance and revealed a much increased ($\sim$ ten times greater) intensity at a closer distance ($\Delta r \sim$ 10,000 km) to the earth than was the case during the flight of *Lunik I* on January 2, 1959.

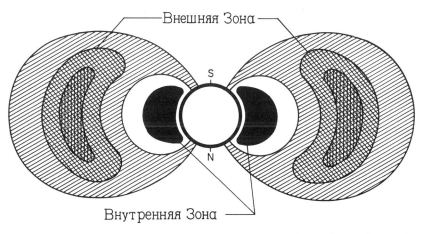

Fig. 35. *A diagram of the structure of the radiation belts of the earth as published in the Soviet newspaper* Pravda *on March 6, 1959.*

The important measurements of ionized gas and fast electrons in the outer magnetosphere by Gringauz et al. [1961a] using ion traps on *Lunik II* are referred to in chapter 11.

A related investigation on *Luniks I* and *II* was the measurement of the magnetic field by Dolginov et al. [1961]. During the traversal of the outer radiation belt in both cases, the observed magnetic field departed (negatively) from the calculated dipolar value in the vicinity of a radial distance of about 20,000 km by several hundred gammas. This departure was attributed to a terrestrial ring current.

On April 22, 1960, Sergei N. Vernov, Alexander E. Chudakov, Nicholas V. Puskkov, and Shmaya S. Dolginov were awarded Lenin Prizes for "discovery and investigation of outer-space radiation of the earth and investigation of the earth's and the moon's magnetic field." The omission of K. I. Gringauz and V. I. Krassovskii is noteworthy.

XI.

Second Generation Investigations and Advances in Physical Interpretation

In April 1958 we conclusively identified the intense radiation observed by *Explorers I* and *III* as electrically charged particles trapped in the external geomagnetic field. This work, together with American and Soviet work during the remainder of 1958 and early 1959, provided a survey of the spatial distribution of the radiation and a crude characterization of its penetrability and energy flux.

But much of the pioneering work, both American and Soviet, was deficient in two important respects: (*a*) the failure conclusively to identify the species of particles responsible for the responses of the various detectors and (*b*) the restriction of the measurements to radiation (either primary or secondary) capable of penetrating $\gtrsim 1$ g cm^{-2} of material. The early work was, of course, deficient in these respects as well as in dynamic range for the simple reason that its objective was to measure the global distribution of the primary cosmic radiation, whose composition, energy, and intensity were already known approximately. Even after recognition of the presence of the trapped radiation, it was difficult to overcome the above-mentioned deficiencies, both because of severe restraints on mass, power, size, and telemetry capacity and because of the rapid time scale of the follow-on investigations (which precluded extensive developmental work). Noteworthy exceptions were (*a*) the work of Krassovskii et al. [1960, 1961], in discovering electrons having energy of ~ 10 keV by very thinly shielded fluorescent screens on *Sputnik III* and (*b*) the work of Gringauz et al. [1961a] using ion traps on the second cosmic rocket (*Lunik II*).

The latter body of work was strongly against a bremsstrahlung interpretation of *Pioneer III* and *IV* data in the outer radiation belt—one of several possible lines of interpretation that I had proposed and the one analogous to our findings for the auroral soft radiation. Gringauz et al. offered the refutation as follows:

The small magnitude of the currents in the +15-v trap over the first part of the trajectory can only be interpreted by assuming that the flux of electrons with energies in excess of 200 ev in the region of the outer radiation belt does not exceed $2 \cdot 10^7$ (cm² sec)$^{-1}$. This result is in direct disagreement with the suggestion that the flux of electrons with energies E = 20–30 kev is larger in the region of maximum intensity of the outer radiation belt. The observed counting rate in experiments described in [Van Allen and Frank 1959a, b; Vernov et al. 1959b, 1961] was explained there by the production of X rays in the body of the container and the screens of the counters due to the impact of soft electrons (E = 20–30 kev). We suggest that the observed counting rate should be explained by a considerably smaller flux of much harder electrons. Associated results are that in the area of maximum radiation of the outer radiation belt the electron kinetic energy density should be smaller by several orders of magnitude than the energy of the geomagnetic field, and that the region of minimum intensity of the magnetic field which has been found at 14,000 km [Dolginov et al. 1961] is apparently unrelated to the diamagnetism of energetic electrons in the radiation belt.

The pure bremsstrahlung interpretation of the counting rates of the Geiger tubes on *Pioneer IV* required in the heart of the outer zone at an altitude of about 16,000 km an omnidirectional intensity of (nonpenetrating) electrons $E_e > 20$ keV of 1×10^{11} (cm² sec)$^{-1}$ and an omnidirectional intensity of electrons $E_e > 200$ keV of less than 1×10^8 (cm² sec)$^{-1}$. Alternatively, the upper limit on the omnidirectional intensity of penetrating electrons ($E_e > 2.5$ MeV) was 1×10^6 (cm² sec)$^{-1}$, and the upper limit on penetrating protons ($E_p > 60$ MeV) was 1×10^2 (cm² sec)$^{-1}$ [Van Allen and Frank 1959a]. In 1960 the work of Gringauz et al. was known in the United States in only rather skimpy and relatively unconvincing form, principally because of the American-Russian language barrier and the long delay in the delivery and translation of Soviet journals.

Another major result in the above-cited paper of the Soviet workers was the determination of the number density of low-energy electrons along *Lunik II*'s trajectory. They found number densities of ~5 × 10² cm^{-3} (for assumed Maxwellian distributions with temperatures in the range 10,000–50,000°K) in the altitude range 1,000–15,000 km and a fall-off by a factor of at least 20 in the range 15,000–20,000 km to below their threshold of detection. This strong fall-off in number density at a radial distance of $\approx 4R_E$ constituted the first direct determination of what was later called the plasmapause (fig. 36). Retrospectively, this work and that of Krassovskii et al. were less influential in American work than they deserved to be, principally because of the sociological-cultural gap between the American and Soviet workers.

A significant theoretical objection to the pure bremsstrahlung interpretation was also raised by Dessler and Vestine [1960], applying a basic theorem of Dessler and Parker [1959]. The latter authors had shown theoretically that the ratio of the total kinetic energy of trapped charged particles to the total volume integral of magnetostatic field energy external to the earth must be approximately equal to the ratio of the consequent (diamagnetic) depression of the equatorial surface field to the vacuum value of that field. During large magnetic storms the observed value of the latter ratio was known to be of the order of 300/30,000 or ~1%. On the reasonable assumption that this fluctuational ratio is of the same order of magnitude as the total value of the ratio, Dessler and Vestine concluded that the particle intensities required by the pure bremsstrahlung interpretation, if present throughout an assumed region of reasonable dimensions, were implausibly high and that the responses of my detectors must have been caused by penetrating electrons and/or penetrating protons.

I had no illusions about the speculative nature of the bremsstrahlung interpretation as stated in the original paper. But as an experimentalist, my principal reaction to these refutations was to redouble my efforts to determine the electron and proton spectra in the outer belt, as well as the inner belt, by observations of a direct and unambiguous nature. This point of view was expressed in the concluding paragraphs of my August 1961 invited discourse to the International Astronomical Union [Van Allen 1962]:

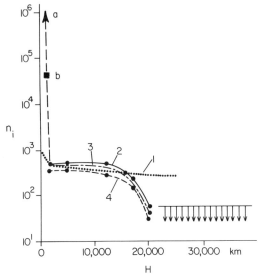

Reprinted with permission from *Artificial Earth Satellites*, vol. 6, K. I. Gringauz, V. G. Kurt, V. G. Moroz, and I. S. Shklovoskii, "Ionized Gas and Fast Electrons in the Earth's Neighborhood and Interplanetary Space," New York: Plenum Press, © 1961.

Fig. 36. Number density of quasi-thermal ions as a function of altitude H above the surface of the earth, observed with Lunik II *[Gringauz et al. 1961a]. The curves labeled 1, 2, 3, and 4 correspond to various interpretations. Points a and b are from observations in the ionosphere at altitudes of 470 and 800 km, respectively, with* Sputnik III *[Gringauz, Bezrukikh, and Ozerov 1961]. One of the most significant features of these results is the discovery of the rapid decline of number density at an altitude of about 20,000 km, later called the plasmapause.*

Moreover even direct observational knowledge of the absolute intensities and energy spectra of electrons and protons in the outer zone is in a quite preliminary state. On the basis of the single assumption that the intensity of electrons of energies exceeding 2.2 Mev does not exceed 10^{-6} of those of lesser energy, the author has given the experimentally based estimate of 10^{11} $(cm^2\ sec)^{-1}$ as the omni-directional intensity of electrons of energy exceeding 40 kev in the heart of the outer zone on 3 March 1959 (a date of exceptionally high intensity); and in spite of considerable later evidence some of a confirmatory and some of a conflicting nature (see for example collection of papers in *Space Research*, edited by H. Kallmann Bijl, 1960), he still finds it very difficult to accept a figure less than about 10^{10} $(cm^2\ sec)^{-1}$ as typifying the intensity of electrons in the tens to hundreds of kilo-electronvolt energy range in the heart of the outer zone.

There remains a pressing need for more decisive experiments in the area. Such experiments are currently underway.

The investigations "currently underway" were represented by the Iowa apparatus on *Explorer 12*, which was launched on August 15, 1961, into a low inclination (33°) orbit having apogee at $12R_E$. Our small magnetic electron spectrometer with thin-window Geiger tube detectors showed that "the omnidirectional intensity of electrons of energy greater than 40 keV [in the heart of the outer radiation belt] is typically of order 10^8 $(cm^2\ sec)^{-1}$ and of energy between 1.6 and 5 MeV, 2×10^5 $(cm^2\ sec)^{-1}$. Hence, the response of lightly shielded (~ 1 g cm^{-2}) detectors is largely due to direct penetration of the primary electrons and [one of] our 1959 assumptions for the tentative interpretation of *Pioneer III* and *IV* observations in the outer zone [is] seen to be invalid" [O'Brien et al. 1962a]. This finding corresponded to the original "alternative" interpretation of *Pioneer III* and *IV* data by Van Allen and Frank [1959a,b]. And it was thoroughly confirmed by the Iowa group by a comprehensive study of *Explorer 12* data [Rosser et al. 1962].

Meanwhile, knowledge of the composition of the trapped radiation in the inner radiation belt was advanced decisively by the flight of a pack of photographic emulsions on a Thor-Able ballistic missile from Cape Canaveral [Freden and White 1959, 1960]. The rocket reached a "maximum altitude of 1230 km and spent about 15 minutes above 1000 km between 20° and 3°N," thus penetrating well into the inner radiation belt. The exposed emulsions were recovered, developed, and found to be crisscrossed by a very large number of particle tracks. The effective shielding of the emulsion totaled about 6 g cm^{-2} of material corresponding to the range of 75 MeV protons or 12 MeV electrons. By track-density identification it was found that over 99% of the tracks were caused by *protons*. The energy spectrum of the protons was determined in the range 75 to 700 MeV. One fit to the data was of the form $N = N_1 E^{-n}$ with E the kinetic energy, N the absolute number of protons (MeV ster sec cm^2)$^{-1}$, $n = 1.84 \pm 0.08$, and $N_1 = 2.1 \mid ^{+1.0}_{-0.7} \mid \times 10^3$, if the entire exposure were attributed to the portion of the trajectory above 1,200 km altitude. Extrapolating this spectrum to lower energies one finds that the absolute intensity of protons $E_p > 30$ MeV is in plausible agreement with the value inferred from the entire counting rate of the less shielded Geiger tube on *Explorer IV*, if the latter rate is averaged over the portion of the Thor-Able trajectory above 1200 km.

In the summer of 1959 I reviewed [Van Allen 1959b] the then available information from *Explorer IV*, *Pioneer III*, *Pioneer IV*, and *Sputnik III* and from the rocket flights of Allen et al. [1959], Freden and White [1959], and Holly and Johnson [1960, shown to me in private communication in 1959] and gave the following summary:

Sample tentative intensities—The following results represent preliminary intensity values in the *heart of the inner zone* at an altitude of 3600 km on the geomagnetic equator:

(a) Electrons of energy greater than 20 kev: maximum unidirectional intensity $\sim 2 \times 10^9$ (cm² sec ster)$^{-1}$.

(b) Electrons of energy greater than 600 kev: maximum unidirectional intensity $\sim 1 \times 10^7$ (cm² sec ster)$^{-1}$.

(c) Protons of energy greater than 40 Mev: omnidirectional intensity $\sim 2 \times 10^4$ (cm² sec)$^{-1}$.

Later in 1959 and during 1960 other rocket flights of nuclear emulsions and of electronic detectors added greatly to knowledge of the composition and intensity of the radiations in the inner zone [Yagoda 1960, Armstrong et al. 1961; Naugle and Kniffen 1961, 1963]. The work of Naugle and Kniffen was noteworthy in that it measured the positional dependence of proton intensity by rotating stacks of emulsions past a port in a heavy shield.

The results of Freden and White (later confirmed in many different ways) were influential in establishing cosmic-ray neutron albedo decay protons as an important and perhaps dominant source of the penetrating particles in the inner radiation belt. Such a source, resulting from the in-flight decay of neutrons emerging from cosmic-ray produced nuclear disintegrations in the atmosphere, is inevitable, of course, but the issues requiring quantitative model calculations are the energy spectrum, the positional and angular distributions of intensity, and, most important, the residence times or lifetimes of trapped protons and electrons. An elementary estimate assures that the absolute intensity of high-energy protons from cosmic-ray neutron albedo decay is of the proper order of magnitude if the typical residence time of protons $E_p \gtrsim 30$ MeV is of the order of tens of years [Van Allen, McIlwain, and Ludwig 1959a].

In connection with planning the Argus tests in late 1957, Christofilos [1959] had thought of the possibility that cosmic-ray neutron albedo might inject a significant number of energetic electrons into durably trapped orbits in the earth's external magnetic field and that the accumulated intensity of such natural electrons might make it difficult or impossible to observe the electrons to be injected artificially by fission decay products from small nuclear bombs ($\sim$ 1 kiloton yield), despite the quite different energy spectra of the respective classes of electrons. I first learned of Christofilos's suggestion of the possibility of a natural radiation belt from Pickering in April 1958, while we were analyzing the data from *Explorers I* and *III*. He was thinking of the β-decay electrons (upper energy limit 783 keV) from thermal neutrons, moving upward from the top of the atmosphere. This idea became widespread later in 1958 and was extended to include the possibility of the injection of high-energy ($E_p \gtrsim$ tens of MeV) protons into trapped orbits upon the decay of neutrons of such energies. The latter possibility was a part of my thinking, as I noted in a telephone conversation with McIlwain on August 8, 1958 (q.v. chapter 8) about detector modifications for *Explorer V*. The general idea of a neutron albedo source of trapped particles was mentioned by Vernov in his August 1958 CSAGI lecture. It was not clear, however, whether he was thinking of thermal or high-energy neutrons as source particles, and I have not been able to find any published record of his having dealt with the idea in a quantitative way at that time. Meanwhile, during the summer of 1958 Paul J. Kellogg and S. Fred Singer were independently undertaking detailed model calculations. The former author credits Thomas Gold and Pamela Rothwell for the central idea of his work.

In mid-August 1958 Singer submitted two short papers to *Physical Review Letters*; both were published in the September 1, 1958, issue of that journal. In the first of these papers [1958a] he proposed that the observed radiation was "trapped cosmic ray albedo." Although Singer did not make it clear in this paper, the "albedo" in question apparently consisted of upward-moving charged particles (principally mesons, protons, and electrons) that originated in cosmic-ray interactions in the upper atmosphere. Such particles had been a central feature of early discussions [Van Allen 1953; Treiman 1953] of the relative importance of primary and secondary cosmic-ray particles in observations of total particle intensity above the atmosphere. However, it was usually supposed that such particles made only one latitudinal excursion or at most one round trip in latitude along a line of force before being lost into the atmosphere. A generally similar concept had been invoked by E. O. Hulburt in 1928 [Chapman and Bartels 1940] as part of his ultraviolet theory of the aurora. Some small fraction of such albedo particles are scattered so that their pitch angles are changed and their mirror points are moved to both higher and lower altitudes. Such scattering can cause temporary trapping, but, by the same token, the residence times are correspondingly brief. On the whole this suggestion of a direct albedo source for trapped particles was quantitatively quite unpersuasive.

Singer's second September 1958 paper [1958b] considered neutrons from cosmic-ray reactions in the atmosphere as a source for trapped protons of energies ranging up to several hundred MeV and gave a preliminary discussion of the source strength, energy spectrum, distribution in space, and loss processes for such protons. This is one of the two early papers putting forward a quantitative study of what was later called cosmic-ray albedo neutron decay (Crand) as a possible source of high-energy trapped protons in the earth's inner radiation belt.

Kellogg's paper [1959a], which was more comprehensive, dealt with the injection, distribution, and lifetime of neutron-decay electrons as well as low-energy (E_p ~5 MeV) and high-energy (E_p ~50 MeV) protons. This paper was circulated privately in late summer 1958, submitted to the journal *Il Nuovo Cimento* on September 24, 1958, and published in the January 1, 1959, issue.

The Crand source has stood the test of much subsequent work and is generally accepted as the primary one for very energetic protons ($E_p \gtrsim$ tens of MeV) in the inner radiation belt [Hess 1962; White 1973]. There are very low intensities of protons $E_p \gtrsim$ tens of MeV outside of 2.2 R_E. This fact is attributed to a combination of three effects: (*a*) a decrease of the source strength with increasing distance from the top of the atmosphere, (*b*) loss of trapping efficacy with increasing values of ρ (grad B)$_\perp$/B (where ρ is the gyro radius of the proton in a magnetic field B), and (*c*) the radially increasing importance of perturbations by wave-particle interactions in the ambient plasma and of temporal fluctuations in the guiding magnetic field. The computed albedo neutron source strength for high-energy protons is so weak (~1 proton per cm^3 per ten million years) that residence times must be of the order of a decade or more to account for the observed inner-zone intensities.

The situation with respect to the concomitant electrons from neutron decay is much more complex, and there has been no convincing reconciliation of the observed intensity and spectrum of electrons with computed expectations for the Crand source.

The penetrating radiation in the inner zone has a very considerable practical importance to space flights around the earth because of its high intensity (~20 roentgens hr^{-1} under 1 g cm^{-2} of aluminum in the heart of the inner zone) and the large mass of material required for significant shielding. In effect the prolonged flight of men, animals, and radiation-sensitive equipment must be restricted to altitudes below 400 km or above 6,000 km [Van Allen 1960]. Single rapid traversals of the inner zone, as in flights to the moon and planets, are tolerable from a radiation point of view.

The principal scientific interest of the magnetosphere lies in a quite different realm of phenomena. The rich and varied body of knowledge of the aurorae, geomagnetic storms, temporal variations of the intensities of trapped particles (particularly in the outer zone and its outer fringes), and the association of these phenomena with solar activity and the rotation of the sun leaves little doubt that fluctuating solar corpuscular streams are the dominant cause of all of these phenomena. Relatively low energy ($E \lesssim 10$ keV) plasma in the radiation belt regions and, of course, the geomagnetic field itself are the other essential ingredients of the physical system.

The inference that solar corpuscular streams are the cause of geomagnetic storms and unusually active auroral displays was already well established in the 1930s [Chapman and Bartels 1940]. In that epoch it was supposed that such intermittent streams of hot ionized gas were emitted radially outwards from active regions on the sun into the vacuum of interplanetary space at a bulk velocity of 500 to 1,000 km sec^{-1}. By virtue of the rotation of the sun, the instantaneous locus of such a stream has the geometric form in interplanetary space of an archimedean spiral as viewed in an intertial coordinate system—in pictorial analogy with the locus of a stream of water from a garden hose whose nozzle is rotated about an axis perpendicular to its length. It is remarkable that a *continual* flow of solar gas was not suggested at that time. In retrospect, the small but well-known daily variation of the geomagnetic field, the usual occurrence of at least mild fluctuations in the field, and the appearance of visible auroral displays on essentially every clear night in the auroral oval would have justified such a suggestion.

The closest approach to a forecast of the existence of the magnetosphere lay in the Chapman-Ferraro [1931, 1932] theory of geomagnetic storms. These authors supposed that the initial phase of a storm, the brief (a few minutes) increase in magnetic field strength at the earth's surface, was caused by the compressive effect on the field of an arriving stream and that the main phase, a marked depression of the surface field strength lasting a day or more, was caused by a westward flowing ring current in or near the geomagnetic equator at a radial distance of perhaps about 6 R_E. The injection of charged particles into such semidurable, circular orbits (protons moving westward, and electrons eastward) was visualized as being the result of polarization electric fields developed as the stream of ionized solar gas encountered the geomagnetic field. Most of the gas was supposed to flow around the geomagnetic field in such a way that the earth produced an elongated cavity stretched out in the antisolar direction, a view developed more fully later by many authors. A kindred but more sophisticated model encompassing the hypothesis that there was a weak magnetic field in the solar stream was developed by Alfvén [1955], who also visualized the development of the observationally inferred ring current.

In a subsequent paper Singer [1957] was concerned with physical mechanisms of geomagnetic storms. He invoked a mixture of the theoretical ideas of Chapman and Ferraro [Chapman and Bartels 1940] and especially Alfvén [1955]. Following Alfvén, Singer suggested that solar corpuscles, which he took to be typified by protons of energy $E_p \sim 20$ keV and electrons of energy $E_e \sim 10$ eV, were "scattered" and injected somehow into the magnetic field of the earth during the early phase of a magnetic storm. He further supposed that such particles were then left in trapped orbits for estimated time periods of the order of a day to constitute a ring current at approximately six earth radii as the direct cause of the main phase of a magnetic storm. Singer visualized that such temporarily trapped particles moved in helical orbits along magnetic lines of force between northern and southern hemispheres and simultaneously drifted in longitude, protons westward and electrons eastward. All of this was in accord with Størmer's theory of bound motion of electrically charged particles in a dipolar magnetic field. It was a detailed improvement over Chapman's hypothesis that the charged particles responsible for the main phase of a magnetic storm circulated around the earth in circular equatorial orbits— a hypothesis that was more mathematical than physical but had been adopted by Chapman principally for gross visualization. Using Alfvén's [1950] guiding center approximation to Størmerian motion, Singer calculated mean longitudinal drift velocities of guiding centers of protons and electrons of assumed energies. He then calculated an electrical current density by an erroneous method [Spitzer 1952; Akasofu and Chapman 1961] by simply multiplying these drift velocities by an assumed number density of particles and further by the unit of electronic charge. Qualitative features of Singer's discussion resemble later observational findings on the physical nature of the ring current [Frank 1967].

Meanwhile, in 1951 Biermann had observed the ionized tails of comets and concluded that the flow of solar gas was essentially continuous, though of fluctuating magnitude [Biermann and Lüst 1966]. In a series of papers in the late 1950s Parker [1963] supported this evidence with theoretical considerations concerning the supersonic outflow of gas from the solar corona. He called such a continuous outflow the *solar wind*, a name that has now largely supplanted the term *solar corpuscular streams*, though many of the basic concepts remain the same. Other evidence for the continuous presence of plasma in interplanetary space came from the scattering of sunlight and radio waves by electrons in the K-corona of the sun and the fluctuations of cosmic-ray intensity [Rossi 1963].

The first direct measurements of the solar wind were made by Gringauz et al. [1961a] with plasma probes on *Luniks I, II*, and *III* and a 1961 flight toward Venus. They observed, at about their instrumental threshold, a flow of positive ions beyond the earth's magnetosphere of the order of 2×10^8 (cm^2 sec)$^{-1}$, the energy of the ions being much greater than 20 eV; but neither the energy distribution nor direction of flow was determined. A more convincing investigation of the interplanetary plasma was conducted by Bonetti et al. [1963] in 1961 with *Explorer X*. These authors found a plasma flowing in approximately the radial direction away from the sun at a velocity of ~ 300 km sec^{-1} and with a flux density of positive ions (presumably protons) of a few times 10^8 (cm^2 sec)$^{-1}$. Snyder, Neugebauer, and Rao [1963] extended such observations in a much more comprehensive way with an electrostatic spectrometer on *Mariner II* during some four months of its interplanetary flight enroute to and beyond Venus in late 1962. They found that the solar wind was present continually, flowing radially outward from the sun. The measured bulk flow velocity varied from 380 to 700 km sec^{-1}, and the inferred number density varied from about 10 to 35 cm^{-3}. Other significant findings were that the mean flow velocity was independent of heliocentric distance over the range 0.7 to 1.0 astronomical units and that there was a convincing positive correlation between flow velocity and the global geomagnetic activity index $\Sigma\,K_p$. This work put the entire subject on a new level of observational reality and served as a prototype of many subsequent investigations [Dessler 1967].

With the benefit of this brief account of the Chapman-Ferraro and Alfvén theories of geomagnetic storms and of early knowledge of the solar wind, it is now possible to sketch the main ideas of the origin and acceleration of most of the trapped particles in the magnetosphere. For this purpose the mean flow velocity of the solar wind is taken to be 400 km sec^{-1}, the mean number density 5 cm^{-3}, and the temperature of the gas $\sim 10^5$ °K. Relative to the earth, protons have a kinetic energy of ~ 1 keV and electrons, their thermal energy of ~ 10 eV. The solar wind is slowed and deflected as it encounters the outer fringes of the earth's magnetic field, because of the induction of electric currents in the ionized, conducting gas. Partial thermalization occurs; that is, the energy of the ions is partially shared with the electrons by plasma-wave interactions (not by collisions in the elementary sense). The process (or processes) by which this thermalized plasma is injected into the outer fringes of the magnetic field of the earth is not clear but is thought to be caused by turbulent, disordered magnetic and electric fields produced by the fluctuating properties of the solar wind. Once electrically charged particles are injected into quasi-trapped orbits, their energies fluctuate randomly in further response to these fields. A particle that gains energy on the average, while conserving its first adiabatic invariant $\mu = p_\perp^2/B$ (where B is the scalar magnitude of the magnetic field and $p_\perp$ is the component of the particle's momentum perpendicular to the magnetic vector), moves toward greater average values of the magnitude of the magnetic field strength, i.e., inward; one that loses energy moves toward lesser values of the field, i.e., outward [Kellogg 1959b, 1960; Herlofson 1960]. The entire process is visualized as diffusion in phase space with the gradient of phase-space density at constant μ being positive outward. The average effect is a net inward diffusion, with losses into the atmosphere of the earth occurring as the inward diffusion proceeds. Typical values of μ in the solar wind are 200 MeV/gauss for protons and 0.2 MeV/gauss for electrons, but there are large variations in both values, especially for electrons. At a radial distance of 1.5 R_E, the resulting energy of a typical proton is ~ 20 MeV and that of a typical electron, 0.020 MeV, but there are presumably wide variations from these typical values.

On the other hand, low-energy ($\sim$ a few eV, with marked upward fluctuations) plasma from the ionosphere also gains access to trapped orbits by virtue of its thermal energy and populates the magnetosphere to form a "background" of quasi-thermal plasma. The observed number density of such plasma is $\sim$1,000 cm^{-3} out to about 3.5 R_E (the plasmapause), beyond which it diminishes to $\sim$40 cm^{-3} [Gringauz et al. 1961a; Carpenter 1963]. Interior to the plasmapause, the plasma co-rotates with the earth. Beyond about 4 R_E, however, convective forces caused by the general electric field induced across the magnetosphere by motion of the magnetic field of the solar wind across the conducting magnetosphere [Axford and Hines 1961] dominate over the forces that result from the co-rotational electric field, and there is a marked loss of plasma through the boundary of the magnetosphere. The entire physical process is a dynamic, time-variable one, responsive to both small and large variations in solar wind flow.

In the stronger magnetic field interior to $\sim$1.6 R_E, the population of energetic particles is relatively stable. Within the history of available observations only a few substantial variations ($\sim$ a factor of 3) have occurred in this region. The corresponding residence times of energetic particles is, therefore, of the order of many years so far as gross variations are concerned. However, outside of $\sim$2.0 R_E large variations (one or two orders of magnitude) occur frequently. The onset of such variations is usually an impulsive depletion (dumping) on the occasion of a magnetic storm. The residence times of energetic particles in the outer belt are of the order of days to weeks [Van Allen 1964]. Following depletions, the population of energetic particles is restored within times of this magnitude [Pizzella, McIlwain, and Van Allen 1962; Forbush, Pizzella, and Venkatesan 1962]. One of the early observational puzzles of this body of phenomena was the fact that the intensities recovered to about the same value irrespective of the average level of solar wind activity. The general impression was the existence of a "saturation level" of the energetic particle population. Some years later this observational fact received an explanation in a seminal paper of Kennel and Petschek [1966], who developed the theory of the "stable trapping limit" on the intensity of charged particles in the presence of a background of quasi-thermal plasma. The theory involved consideration of wave-particle interactions and conditions for the growth or decay of spontaneous plasma instabilities. There is now an impressive level of agreement between observed maximum particle intensities and theoretically derived values.

A recent specific test [Hovestadt et al. 1978] of the solar-wind origin of energetic ions in the outer radiation belt has received an affirmative answer by comparing elemental abundances of minor constituents of the trapped population with those in the solar wind. Nonetheless, there is evidence that ionospheric ions are the source of a small fraction of the energetic particle population.

It is quite unreasonable that thermalized solar wind protons having an energy ~1 keV in a region of injection into the magnetosphere (at say 10 R_E) can achieve energies of the order of 100 MeV upon diffusing inward to the inner radiation belt. Indeed, it is theoretically impossible if the first adiabatic invariant is conserved. However, there is considerable evidence in favor of the sporadic injection of solar energetic particles ($E_p > 10$ MeV) into trapped orbits at ~3.5 R_E, and such particles (principally protons) may constitute a significant source of inner-zone trapped protons. There is also evidence for the presence of energetic α-particles and C, N, and O ions in the inner zone, probably from this source [Van Allen, Randall, and Krimigis 1970]. Because such heavy ions cannot arise from neutron decay, this evidence argues for a mixed origin of the particles in the inner belt. Even in 1983, there is no completely satisfactory explanation for the very energetic protons in the inner belt, but most students of the field favor the cosmic-ray neutron albedo source as being of primary importance. Part of the reason for this preference, it must be pointed out, is the fact that this process is theoretically more tractable than are other processes such as direct particle injection near the earth. It has not been shown with any persuasion that the energetic electron population of the inner zone can be attributed primarily to the neutron-decay process. The physical dynamics (scattering, diffusion, and energy loss) of trapped electrons are admittedly more complex than those of protons.

Despite the successes of low-altitude satellites, rocket flights, and a few deep-space probes during late 1957, 1958, and early 1959, there was a widely recognized need for detectors that would clearly discriminate among different species of particles (electrons, protons, α-particles, etc.) and measure their separate energy spectra, all as a function of pitch angle and position. In addition and more urgently, there was a need to extend the energy thresholds of detectors to much lower energies, say a few eV, in order to obtain full knowledge of the total population of trapped particles, including the plasma. Also, observations of the steady state and fluctuating electric and magnetic fields throughout the magnetosphere were becoming increasingly important to theoretical understanding of the acceleration, diffusion, and loss of particles. The earliest papers to emphasize the plasma physical nature of magnetospheric dynamics were those of Gold [1959b] and Kellogg [1959b]. In situ observational work on the electrostatic and electromagnetic waves, the complementary part of the wave-particle interaction system, was initiated later by Leiphart [1962] and Gurnett and O'Brien [1964].

The design of particle detectors for satellites depended on a rich heritage of experience within laboratories devoted to low-energy nuclear physics and plasma physics. Adaptation of these techniques to space flight required the solution of difficult problems arising from the necessity for miniaturization, low mass, low power drain, limited telemetry capacity, large dynamic range, mechanical ruggedness, durability, and proper operability over a large range of temperatures. Moreover, the particle "beam" in the magnetosphere is more-or-less isotropic and of mixed and initially unknown composition, in contrast to the usually collimated beams of particles in laboratory nuclear physics, and the investigator has no control whatever over the intensity or other parameters of the radiation to which his detectors are subjected. It is perhaps then no surprise that space-qualified radiation detectors have been in a continual state of evolutionary development for over two decades, following the use of relatively primitive detectors in early work.

As of early 1959 it was also clear that interplanetary observations and greater positional and temporal coverage of the magnetosphere were urgently needed in order to learn the "full picture," including the study of time variations and the cause-and-effect relationships with solar and geomagnetic activity. Three types of investigations were required:

1. Ones with long-lived satellites in relatively low-altitude ($\sim 1,000$ km) circular orbits steeply inclined ($\sim 90°$) to the equator for study of the lower fringes of both inner- and outer-radiation belts, precipitation of particles (especially in the auroral zones), relationship of the aurorae to the outer belt, the high latitude limit of trapping, and the arrival of solar energetic particles over the polar caps.
2. Ones with long-lived satellites in quite elliptical orbits (apogees beyond at least 15 R_E) for more comprehensive investigation of the structure and temporal variations of the outer belt and the outer fringes of the magnetosphere.
3. Ones with satellites in extremely high eccentricity earth orbits (apogees beyond 30 R_E) or lunar orbiters for monitoring solar corpuscular streams in interplanetary space.

In a more expansive vein of thought many of us were also urging that magnetospheric investigations of the other planets of the solar system be undertaken as soon as they were technically feasible. The preliminary measurements of the magnetic field of the moon by *Lunik II* in September 1959 showed that the moon has little or no general magnetic field [Dolginov, Zhuzgov, and Pushkov 1960].

The first effort to make energetic particle and magnetic-field observations with a satellite in a markedly eccentric orbit was represented by *Explorer VI*, a NASA satellite built by the Space Technology Laboratories and launched on August 7, 1959. The initial orbit of this solar-battery-powered satellite had an inclination of 47° to the earth's equator (38° to the ecliptic) with perigee at an altitude of 237 km and apogee at a radial distance of 7.62 R_E. Useful scientific data were obtained for about two months. Radiation detectors comprised an integrating ionization chamber and single Geiger tube of the University of Minnesota [Arnoldy, Hoffman, and Winckler 1960b,c, 1962; Arnoldy et al. 1962; Hoffman 1961; Hoffman, Arnoldy, and Winckler 1962a,b; Winckler, Bhavsar, and Anderson

1962], a shielded array of seven semiproportional counters of the University of Chicago [Fan, Meyer, and Simpson 1960d, 1961], and a thinly shielded plastic scintillator of the STL [Rosen, Farley, and Sonett 1960; Rosen and Farley 1961]. Advances were made in the separate identification of energetic protons and electrons and their absolute intensities, in the knowledge of the positional structure of the outer radiation belt, and especially in the large temporal changes of intensities associated with geomagnetic storms. A magnetometer, also prepared by STL, observed the magnetic signature of an extraterrestrial current ("ring" current) system centered at a geocentric distance of about 6 R_E.[Sonett, Smith, and Sims 1960; Sonett et al. 1960a,b].

The University of Iowa's radiation satellite *S46* was launched on March 23, 1960. It was intended for an eccentric orbit with apogee at about 4 R_E, but the Juno II launch vehicle did not produce an orbit.

Meanwhile, *Explorer VII* was successfully launched on October 13, 1959 into a near-circular orbit (altitude range 560 to 1,100 km) inclined 51° to the equator. This NASA/ABMA satellite, also solar battery powered, yielded a large body of synoptic data on the intensity of energetic particles in both outer- and inner-radiation belts until February 17, 1961, from the University of Iowa instrument [Ludwig and Whelpley 1960]. There were intensity variations of as much as two orders of magnitude in the low altitude horns of the outer belt [Forbush, Venkatesan, and McIlwain 1961; Forbush, Pizzella, and Venkatesan 1962], in correlation with geomagnetic storms and with variations observed at large radial distances by *Explorer VI*. Variations by factors as large as 3 were also observed in the lower part of the inner belt again in correlation with solar/geomagnetic events [Pizzella, McIlwain, and Van Allen 1962]. In addition *Explorer VII* provided the first comprehensive survey of solar energetic particle events (over the north and south polar caps of the earth) and showed how common such events were. In the sixteen-month period, October 1959–February 1961, twenty-one distinct events were observed, including the event of April 1–2, 1960 [Van Allen and Lin 1960; Chinburg 1960], also observed by *Pioneer V* remote from the earth [Fan, Meyer, and Simpson 1960a,b], and the great events of mid-November 1960 as reported in the comprehensive *Explorer VII* survey of Lin and Van Allen [1964]. Finally, *Explorer VII* fulfilled our originally proposed objective of *Explorer I* by providing a comprehensive latitude-longitude survey of the cosmic-ray intensity above the atmosphere [Lin, Venkatesan, and Van Allen 1963] (fig. 37). It was found that the intensity increased by a factor of 3.6 from the geomagnetic equator to an invariant latitude of 52° and was accurately constant in the latitude range 52° to 64°. The position of the high-latitude "knee" of the intensity versus latitude curve showed that cosmic rays having magnetic rigidities less than 1.7 GV (corresponding kinetic energy of protons, 1.0 GeV) were absent. These results referred to the period between October 1957–February 1961 (soon after solar maximum activity).

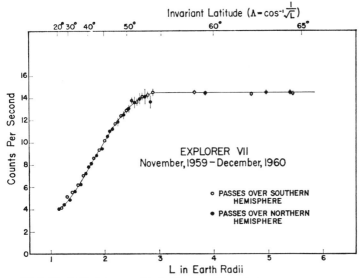

W. C. Lin, D. Venkatesan, and J. A. Van Allen, "Latitude Survey of Cosmic-Ray Intensity by Explorer 7, October 1959 to February 1961," *Journal of Geophysical Research*, vol. 68 (1963), p. 4980, copyrighted by the American Geophysical Union.

Fig. 37. Dependence of the cosmic-ray intensity above the atmosphere on the McIlwain shell parameter L or the invariant latitude Λ as derived from an extensive body of observations by Explorer VII *in both northern and southern hemispheres [Lin, Venkatesan, and Van Allen 1963].*

Our previous rockoon flights during the early 1950s (near solar minimum activity) gave a latitude factor of nine and a high latitude knee at about 56° [Meredith, Van Allen, and Gottlieb 1955], showing the greater prevalence of low-energy cosmic rays during periods of minimum solar activity.

Explorer X was launched on March 25, 1961, into an orbit inclined at 34° to the earth's equator with perigee at ~160 km and apogee at a radial distance of 46.6 R_E. The orbit itself was a durable one, but, because electrical power for the satellite's electronics was supplied by chemical batteries, data were obtained for only about 52 hours, or nearly to the time of first apogee passage. The scientific instrumentation included a plasma probe (retarding potential analyzer) for measuring low-energy protons (0 to 2,300 eV), the first such U.S. device to be flown in space; two fluxgate, single-axis magnetometers; and a rubidium-vapor magnetometer, arranged with a field producing bias coil to yield vector magnetic field measurements. Many fresh results were obtained, including convincing detection of the solar wind [Bonetti et al. 1963] (see previous section) and a wealth of magnetic field data [Heppner et al. 1963]. The combination of the two bodies of data defined the evening-side boundary of the previously hypothetical magnetospheric "cavity" surrounding the earth. Such a cavity within the solar wind was first envisioned by Chapman and Ferraro [1931, 1932]. More recent treatments had been given by Parker [1958], Piddington [1960, 1969], Beard [1960], Johnson [1960], Dungey [1961], Axford [1962], Kellogg [1962], Spreiter and Briggs [1962], and many others.

The next radiation belt satellite to be placed in a highly eccentric orbit was NASA's *Explorer XII*, launched on August 16, 1961. It carried (*a*) a three-axis flux-gate magnetometer (University of New Hampshire); (*b*) a 4.4 mg cm^{-2} thick ZnS (Ag) phosphor on the face of a photomultiplier tube with an ingenious, rotatable absorber wheel and a particle scatterer for separating protons from electrons and for measuring the spectrum of protons (Goddard Space Flight Center); and (*c*) a small magnetic spectrometer equipped with thin-window (1.2 mg cm^{-2}) Geiger tube detectors for separating electrons from protons and for measuring the spectrum of electrons, a shielded Geiger tube similar to ones flown on *Explorers IV, VI*, and *VII* and on *Pioneers III, IV*, and *V*, and three crystals of CdS for detection of the integrated energy flux of charged particles (University of Iowa). The orbit of *Explorer XII* was inclined 33° to the earth's equator with perigee at an altitude of 300 km and apogee at a radial distance of 13 R_E. Initially, apogee was near local noon (in contrast to the orbit of *Explorer X*, whose apogee was at a local time of about 23 hours). Observations were obtained for 112 orbits from launch until December 6, 1961.

The electron data from the magnetic spectrometer and other detectors on *Explorer XII* conclusively resolved the ambiguity of interpretation of our earlier (*Pioneer III / IV*) shielded Geiger tube data. It was found that the intensity of electrons of energy greater than 40 keV in the heart of the outer zone was of the order of 1,000 times less than that required by the pure bremsstrahlung interpretation and that the response of an omnidirectionally shielded Geiger tube there was caused primarily by directly penetrating electrons [O'Brien et al. 1962a; Rosser et al. 1962]. Measurements with an improved system of detectors in the Iowa instrument on *Explorer XIV* confirmed this general result and yielded typical maximum omnidirectional intensities of electrons in the heart of the outer zone as follows:

$E_e > 40$ keV: $\sim 1 \times 10^8$ (cm^2 sec)$^{-1}$
$E_e > 230$ keV: $\lesssim 1.5 \times 10^6$
$E_e > 1.6$ MeV: $\sim 2 \times 10^5$

There were large fluctuations from pass to pass, mostly to lesser values [Frank, Van Allen, and Macagno 1963; Frank et al. 1963; Frank, Van Allen, and Hills 1964]. In the complementary study of protons in the outer zone, Davis and Williamson [1963] found that the maximum unidirectional intensity of protons in the energy range greater than 100 keV was typically 6×10^7 (cm^2 sec sr)$^{-1}$ at 3.5 R_E, and they measured the proton spectrum as a function of radial distance. They also showed that "the total kinetic energy of the trapped low energy protons is greater than that of any other known population of trapped particles and thus their disturbance of the geomagnetic field is greatest" and suggested that such protons might be the principal cause of Chapman's extraterrestrial ring current.

Other major results of Explorer XII were Cahill and Amazeen's [1963] observation of a sharp decrease in the magnetic field strength by a factor of two at a radial distance of 8 to 10 R_E near the noon meridian. This discontinuity was later named the "magnetopause," the outer boundary of the magnetosphere. Confirmatory evidence of the physical nature of the magnetopause was provided by the simultaneous discovery by Freeman, Van Allen, and Cahill [1963] of a region of high particle energy flux (later called the magnetosheath) immediately outside the magnetic discontinuity and of several R_E radial extent. This effect was attributed to the pile-up of a body of quasi-thermalized solar wind gas that acts as a pressure-transmitting "cushion" between the directed flow of the solar wind, which ceases at the hypersonic bow shock, and the geomagnetic field barrier. In a comprehensive reduction of *Explorer XII* data for the period between August 14 and October 31, 1962, Freeman [1964] found that the average radial distance to the sunward magnetopause was 66,000 km (10.3 R_E) but that solar wind fluctuations caused excursions of the stand-off distance from about 51,000 km (8 R_E) to ones beyond the apogee of the satellite's orbit (13 R_E). Large temporal variations in the whole body of particle phenomena occurred from pass to pass, in general correlation with the geomagnetic $D_{ST}(H)$ index.

In addition, quoting Freeman, it is important to note that:

Using data primarily from a system of three-electrode "ion traps" on Lunik 2, Gringauz and collaborators [Gringauz and Rytov 1960; Gringauz et al. 1961b, 1963] discovered omnidirectional intensities $J_o \sim 2 \times 10^8$ (cm^2 sec)$^{-1}$ of electrons of E > 200 ev in the radial distance range from the center of the earth 61,400 to 81,400 Km (55,000 to 75,000 Km above the surface of the earth), and at low latitudes. Gringauz et al. regarded these observations as establishing the existence of a "third radiation belt." Van Allen has estimated [Freeman, Van Allen, and Cahill 1963] that Lunik 2 observations were at a geocentric angle of 125° ($\pm$ 10°) eastward from the subsolar point.

More recently, using space probe Mars 1 data, Gringauz et al. report the observation of the "outermost (third) charged particle belt" at high latitude within *2 to 3 earth* radii. These observations are at a geocentric angle nearly normal to the earth-sun line [Gringauz et al. 1964].

These Soviet findings are concordant with the belief that the CdSTE energy flux response at a westward geocentric angle with the subsolar point of approximately 125° in the region beyond which higher-energy trapped radiation is observed is due predominantly to very soft electrons.

The relationship of this belt of soft electrons to the similar band of soft electrons found on the sunward side of the earth, *outside the magnetosphere*, is not yet clear.

Magnetospheric boundary phenomena were reviewed in 1963 by Frank and Van Allen [1964].

Meanwhile, the Office of Naval Research/University of Iowa satellite *Injun I* (launched on June 29, 1961, into a near-circular, 67° inclination orbit) provided a large body of low-altitude data on absolute electron intensities in both the inner and outer radiation belts, as well as a valuable new body of observations on the high latitude limit of geomagnetic trapping and on the primary auroral particles [O'Brien et al. 1962b; O'Brien and Laughlin 1962; Frank, Van Allen, and Craven 1964].

In accordance with the plan of this monograph (cf. Preface), I decided to conclude it with about the year 1962, following definitive study of the magnetopause and magnetosheath. I have deferred to a later time, however, a review of the magnetospheric consequences of four high-altitude nuclear bomb bursts: the U.S. Starfish burst over Johnston Atoll on July 9, 1962, and three USSR bursts on October 22, October 28, and November 1, 1962. All four of these, especially the first, injected large numbers of energetic electrons into durably trapped orbits in the geomagnetic field (fig. 38). A full account of the consequent observational and theoretical studies is a major feature of magnetospheric literature.

Since the early 1960s there have been major advances in magnetospheric physics, both observational and theoretical. These advances have centered on (*a*) low-energy plasma and associated plasma-wave phenom-

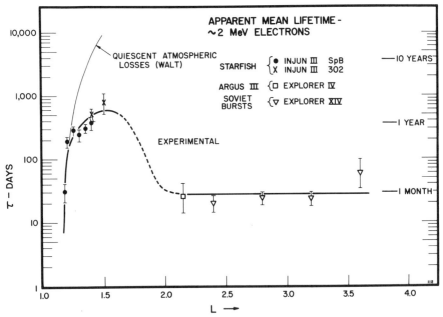

Reprinted by permission from *Nature*, vol. 203, no. 4949, p. 1007. Copyright © 1964 Macmillan Journals Limited.

Fig. 38. The L-dependence of observed values of the apparent mean lifetime of ~ 2 MeV electrons from two U.S. and two USSR nuclear detonations at high altitudes.

ena, (b) the role of large-scale electric fields, (c) the magnetic topology of the distant geomagnetic field, especially in the antisolar side of the earth (the magnetotail), (d) the entrance of solar-wind plasma into the magnetosphere, and (e) the acceleration of auroral particles. Such work continues in a lively way and is yielding a fundamental understanding of the magnetosphere as a very complex plasma physical system—an understanding only vaguely perceived in many respects by 1962.

I conclude with a few remarks on the gross energetics of the magnetosphere. As a start, I assume that the solar wind is the primary energy source for all magnetospheric phenomena. Adopting, as representative values, a number density n of 7 protons cm^{-3} and a flow velocity v of 400 km s^{-1}, one finds a power flux $(nv)(\frac{mv^2}{2})$ of 0.38 erg (cm^2 s)$^{-1}$. The cross-sectional area of the magnetosphere perpendicular to the solar wind flow is that of a circle of approximate radius 14 R_E or 2.5×10^{20} cm^2. Hence, the total power that is potentially available from the solar wind is of the order of 1×10^{20} erg s^{-1} on an average day. During days of high solar activity the power flux increases by as much as an order of magnitude.

In a bright aurora the energy flux of particles into the atmosphere is of the order of 500 erg (cm^2 sec)$^{-1}$ [McIlwain 1960b]. If one approximates the area of each of the two auroral zones by a strip of 2° latitudinal width centered at latitude 65°, the total area of precipitation is 8×10^{16} cm^2. Hence, the instantaneous power required to sustain such an auroral display is 4×10^{19} erg s^{-1}. This value is of the order of 5% of that available under enhanced solar wind flow conditions. The power requirement of the aurorae under relatively quiescent conditions is less than this value by a factor of the order of 50 or about 1% of that available from the quiescent solar wind flow. Thus, it appears reasonable to regard the solar wind as energetically capable of sustaining all known auroral phenomena. Furthermore, the physical coupling between the solar wind and the acceleration and diffusion of charged particles is amply established both theoretically and observationally [Herlofson 1960; Parker 1960; Davis and Chang 1962; Schulz and Lanzerotti 1974].

Another line of thought on this matter is represented by one of my early conjectures [Van Allen 1959b; Van Allen, McIlwain, and Ludwig 1959b] that the geomagnetically trapped radiation is a reservoir of both particles and energy from whose outer fringes the primary auroral particles are precipitated by fluctuations in solar wind flow. This conjecture can be assessed as follows. The total magnetostatic field energy (vacuum case) of the geomagnetic field external to the earth is given by

$$W = \frac{M^2}{3a^3}$$

where M is the earth's dipole moment (8.0×10^{25} gauss cm^3) and a is the radius of the earth (6.378×10^8 cm). Thus,

$$W = 8.2 \times 10^{24} \text{ erg}.$$

By the theorem of Dessler and Parker [1959] the total energetic content of all trapped particles cannot exceed a few percent of W or, say, 2.5×10^{23} erg. This amount of energy can sustain a bright aurorae for about 10^4 sec, *if all particles throughout the magnetosphere are available to be precipitated into the auroral zone*. As soon, however, as it became clear from the Argus tests (chapter 8) that magnetic shells of particles are *not* interchanged in L value on a time scale anywhere nearly this rapidly, the suggestion of simple precipitation of existing particles into the narrow range of latitudes in the auroral zones was recognized as invalid, and it was concluded that the pure dumping conjecture was inadequate energetically by at least three orders of magnitude. The *Explorer IV* and *Explorer VII* data did show a marked depletion of the outer radiation belt on the occasion of large magnetic storms (and exceptional auroral displays), but the depletion was principally at electron energies (~ 1 MeV) far above those characteristic of the auroral primaries (~ 5 keV). Later with *Injun I* and *Explorer XII*, we confirmed the impulsive depletion of ~ 1 MeV electrons *but* found an enhanced intensity of electrons having energies of ~ 40 keV. For these various reasons I abandoned the dumping hypothesis for the connection between aurorae and trapped particles [Van Allen, Lectures on Solar-Terrestrial Physics, 1961], a changed point of view that was further developed and published by my colleague, O'Brien [1962]. There remains little doubt that the aurorae and phenomena of the trapped particle population are closely related, but it appears that neither is the direct cause of the other but that both bodies of phenomena have a common basic cause, the solar wind and the fluctuations thereof. The energy involved in moderately strong magnetic storms is typically of the order of 1×10^{22} erg [Chapman and Bartels 1940], about 1% of the external magnetostatic energy of the earth, a value that is compatible with observed changes in the energetic content of the radiation belt population and directly attributable thereto [Dessler and Parker 1959].

Another conceivable source of energy for magnetospheric phenomena is the rotation of the earth, whose total rotational kinetic energy is 2×10^{36} erg, an enormous value relative to any magnetospheric requirement. But the total magnetostatic energy of the earth (including that in its interior) is only of order 10^{26} erg, and the time scale of the coupling of rotational kinetic energy to magnetic field energy is of order 200,000 years [Cox 1975]. Hence, there is no basis for supposing that the rotational energy of the earth is available for magnetospheric phenomena on a time scale of the order of a few hours or even a few years.

Finally, it may be noted, relative to the Crand source of inner-belt protons, that the total power flow of cosmic rays on the earth is of the order of 2×10^{16} erg s^{-1}. Of this energy, less than 10^{-4} or 2×10^{12} erg s^{-1} goes into trapped particles. Hence, Crand makes a totally trivial contribution to the energetics of magnetospheric processes.

The total power of solar electromagnetic radiation that is absorbed by the earth is about 1×10^{24} erg s^{-1}. A small fraction of this energy sustains the ionosphere, and, in turn, the ionosphere is physically coupled to the magnetosphere by virtue of its conductivity. But the coupling appears to be a quasi-stationary one, and there has been no suggestion of a way in which this coupling can be responsible for significant energization of charged particles.

In summary, it appears reasonably certain that the solar wind is the only significant source of energy for the dynamics of the magnetosphere. Also, it is probably the dominant source of the particles observed therein including those that cause aurorae, though the ionosphere, solar energetic particles, and Crand also contribute to the particle population (fig. 39).

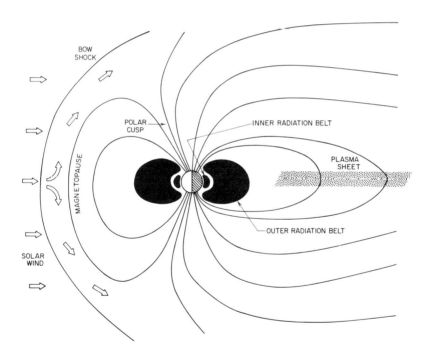

Van Allen, "Radiation Belts," from Lerner/Trigg, ENCYCLOPEDIA OF PHYSICS, © 1981. Addison-Wesley, Reading, MA. p. 832. Reprinted with permission.

Fig. 39. Principal features of the earth's magnetosphere to approximate scale as viewed in cross-section in the noon-midnight meridian plane. The earth is represented by the circle in the central part of the diagram.

APPENDIX A

Proposal for Cosmic Ray Observations in Earth Satellites, 1955

Department of Physics
State University of Iowa
Iowa City, Iowa

28 September 1955

Dr. Joseph Kaplan
National Academy of Sciences
2101 Constitution Avenue
Washington 25, D. C.

Dear Dr. Kaplan:

There is enclosed a "Proposal for Cosmic Ray Observations in Earth Satellites". Recent discussion with Dr. G. F. Schilling has indicated that it is appropriate to submit definite proposals at this time.

I hope that it may soon be possible to establish a committee of interested persons to deal with the diversity of scientific and technical problems involved in the satellite program. Our own proposal is necessarily indefinite in important respects due to the preliminary status of essential aspects of the program.

Best personal regards.

Sincerely yours,

J. A. Van Allen

Department of Physics
State University of Iowa
Iowa City, Iowa

28 September 1955

*Proposal for
Cosmic Ray Observations
in Earth Satellites*

A. General Scientific, Technical and Administrative Considerations

1. *Introduction*

Cosmic ray observations above 50 kilometers altitude have a special simplicity and importance because only above such altitudes can one's apparatus be placed in direct contact with the primary radiation before its profound moderation in the earth's atmosphere. During the past nine years the writer and his colleagues have conducted a comprehensive series of experiments on the primary cosmic radiation by means of V-2 rockets, Aerobee rockets, and most recently a large number of inexpensive balloon-launched rockets.

Because of the special needs of this field of research we have pioneered in the development of inexpensive rockets for physical research at high altitudes (Aerobee and rockoon). And we have pioneered in the conduct of rocket experiments over a wide range of geographical location:

(a) Aerobee expedition to central Pacific in March 1949 (first non–White Sands rocket experiments in any field);

(b) Aerobee expedition to Gulf of Alaska (in January 1950);

(c) Rockoon expedition to Arctic in July 1952 (first Arctic rocket experiments in any field) and again in summers of 1953, 1954 and 1955.

The early work, including the development of the Aerobee, was supported by the Naval Bureau of Ordnance via the Applied Physics Laboratory of the Johns Hopkins University; the work since January 1951 has been supported jointly by the State of Iowa and by the Research Corporation of New York, the U. S. Office of Naval Research and the U. S. Atomic Energy Commission. Recently we have started preparatory work for a series of experiments during the International Geophysical Year 1957–58 with help from the National Academy of Sciences/National Science Foundation program for the I.G.Y.

A bibliography of our completed work in this field is appended. A prospectus of proposed I.G.Y. experiments with "conventional" rockets is on file at the N.S.F. offices for the I.G.Y. and at the National Academy Offices.

2. Proposed Observations (Long Range Planning)

A satellite of the earth has several distinct advantages over conventional rockets as a vehicle for cosmic ray apparatus. These advantages are detailed below in terms of the experiments which are proposed:

(a) The satellite, if in a pole to pole orbit, will complete a comprehensive world-wide geographical survey in a few days. A comparable world-wide survey with conventional rockets would require several hundred flights. Such surveys are essential to learning the geographical distribution of arriving cosmic radiation. The information is first of all a matter of basic geophysical interest. Secondly, by means of geomagnetic theory, it is possible to interpret the data from such a survey into composition, absolute intensities, and energy spectra which characterize the primary cosmic radiation. These deduced data are fundamental to understanding the astrophysical character of the radiation (origin, propagation, and nature of medium through which it passes from source to earth) and the great wealth of high energy nuclear phenomena which occur in the earth's atmosphere. Comprehensive geographical surveys of this nature have been one of our prime objectives for the past nine years. It has been proposed to continue them by conventional means during the next several years as well. However, a successful satellite apparatus would provide a far more efficient method of conducting such surveys. Specifically, a week of satisfactory data will supplant some twelve years of work. Moreover, occasional satellite flights distributed over a period of some years will make possible a systematic record of long term temporal changes. (Ideally, of course, a "permanent" satellite would be desirable).

(b) A subsidiary result of the satellite survey contemplated in (a) will be a mapping of the effective geomagnetic field. This will be done by harmonic analysis of the cosmic ray observations in such a way as to determine the axis of symmetry of the effective, long range magnetic field of the earth. A pole-to-pole orbit is the most desirable for this purpose; but an orbit in approximately the plane of the geographical equator of the earth would also serve the purpose since the geomagnetic axis is inclined to the geographical axis. A knowledge of effective geomagnetic coordinates (for cosmic ray purposes) is essential to the interpretation of a wide variety of ground and balloon observations. The approach proposed is a novel one.

(c) An apparatus continuously orbiting above the earth's atmosphere is ideally situated for studying the correlation of fluctuations of cosmic ray intensity with terrestrial magnetic and solar activity. Such observations should provide important information on the magnitude and nature of the portion of the primary radiation which has a solar origin. And it should clarify the nature of the astrophysical environment of the earth.

(d) Many desirable measurements on the primary cosmic radiation are impractical with conventional rockets because of the brief time such rockets spend above 50 kilometers. The principal class of such intrinsically low intensity measurements is that having to do with establishing the Z-spectrum of the pure primary beam (i.e., the distribution of nuclear species in the primary radiation before its encounter with the atmosphere). A leading astrophysical question is that of the relative cosmic ray abundance of the light elements H/He/Li/Be/B/

C/N/O/F. This laboratory has recently developed a scheme for such measurements using Cerenkov detectors. The operation of such equipment has been successfully demonstrated in Skyhook balloon flights to 98,000 ft. The information can be transmitted by radio telemetering. This apparatus is, therefore, adaptable to use in a satellite (in contrast to the photographic emulsion method which requires physical recovery of equipment). A satellite orbit in approximately the equatorial plane is the most desirable for initial work on this aspect of the primary radiation in order that the geomagnetic latitude remain approximately constant during the course of the measurements. Use of pole-to-pole orbits should follow as a later development.

(e) By means of simple measurements extending to distances from the earth's surface comparable to the earth's radius (preferably several earth's radii) a new attack can be made on the importance and nature of the cosmic ray albedo of the earth's atmosphere. Such albedo consists of products of nuclear reactions in the upper levels of the atmosphere which happen to proceed in upward directions. Any such products which are electrically charged tend to be confined to the vicinity of the earth by its magnetic field in a way which is dependent on geomagnetic latitude. Measurements of the total cosmic ray intensity as a function of distance from the earth will provide a direct determination of the magnitude and nature of the albedo. The results are of general geophysical interest; and more importantly they will serve to improve the knowledge of the correction of the intensity of apparent primary particles of nuclear charge $Z = 1$.

(f) A suitable detector in a satellite should provide a powerful new means for charting the arrival of auroral radiations at the top of the earth's atmosphere. Work of this sort is presently being actively pursued by this laboratory since its discovery of such radiation by direct detection in rockoon flights in the Arctic in 1953. The necessary apparatus is simple in nature and similar to that used in the cosmic ray observations. An approximately pole-to-pole orbit is necessary.

3. *General Technical and Administrative Considerations*

(a) For the purposes of the present proposal it is assumed that "the satellite vehicle" will be developed in stages of increasing technical difficulty so that the final orbiting object will be successively (in something like the following sequence):
(1) an inert, non instrumented object of diameter about two feet and weight about five pounds
(2) an instrumented object of diameter about three feet and weight about fifty pounds
(3) more massive objects.
It is the second stage of the development in which we are principally interested. In fact the first stage may be of such limited utility as to make it advisable to bypass it entirely. However, we would be interested in using something intermediate between (1) and (2). A spherical shape is not important for our uses but a cylinder should have a diameter of at least six inches.

(b) It is further assumed that a satellite program which is to serve a really important scientific function will be a continuing one in which there will be many satellites flown over an indefinitely extended period of time.

(c) It is further supposed that the assignment of payload space in successive satellites will be done by a technical committee of broad interest and competence.

(d) In order that the development of the vehicles themselves be intelligently planned it is of upmost importance that the needs and desires of those contemplating use of the vehicles for scientific work be adequately taken into account in connection with all major technical decisions (i.e., such as size, payload, weight, configuration, choice of orbit, etc.).

(e) Satisfactory telemetering communication from a satellite and satisfactory tracking are problems of major importance. And at least the first depends upon development of suitable sources of circuit power. Moreover, the establishment of an adequate network of ground observing stations is an undertaking comparable to that of constructing the vehicles themselves. Again, cooperative effort of all interested laboratories is essential to a useful final result. A good foundation for all of these developments already exists. But special technical problems must be solved.

(f) Development of individual pieces of scientific apparatus is presumably best done by individual participating laboratories. The design of such apparatus is, however, impossible without detailed knowledge of power supply, telemetering and tracking developments, characteristics, and capabilities.

B. Specific Proposal for Initial Work

1. *Pole-to-Pole Orbit*

The first objective in this case will be to construct an apparatus for measuring the total cosmic ray intensity in absolute units—see paragraphs A2(a), (b), (c) and (f) for purposes. Two types of detectors are under consideration—a single Geiger counter and a Na I (Tl) scintillator. We have extensive experience with both types in high-g rocket apparatus. In either case the system will be designed to transmit counting rate data. If the satellite is to be in continuous telemetering communication with ground stations the usual methods of data transmission will be used. If it will be in intermittent communication, a coded integrating system will be used. A fifty pound payload will be adequate for one week's operation, using existing types of conventional batteries for circuit power. A number of prove-in flights in conventional balloons and rockets will be made. Substantial reduction in weight will be possible if satisfactory solar batteries are available.

2. *Equatorial Orbit*

In this case apparatus of the type described in paragraph B1(a) will probably be prepared first. This will serve purposes A2(b) and in part A2(c). Adequate experience with apparatus involving Cerenkov detectors is also available in this laboratory. Hence apparatus suitable for studying the Z-spectrum of the primary radiation—purpose A2(d)—will also be prepared and proved-in in the initial work. Overall payload weight will not exceed 50 pounds, and will be about 30 pounds if satisfactory solar batteries are available.

3. *Time Scale and Scope of Work*

Preparatory work can start immediately. Final design and construction of actual flight apparatus can be completed in about one year if telemetering system is available by then. We do not propose to develop the telemetering system but should like to participate in choice of characteristics and in preliminary testing.

The specific proposal contemplates construction of some twenty sets of apparatus—of which about five will be flown in satellites, the others being expended in preliminary tests.

The duration of the proposed initial work is approximately three years—January 1956 to December 1958.

C. Proposed Budget for Initial Program

1. *Principal Investigator*, salary from other funds

2. *Post-Doctoral Research Associate* for two years	$12,000
3. Two or three part time *graduate research assistants* for three years	13,500
4. Equipment and Expendable supplies including preliminary test vehicles (telemetering and tracking apparatus assumed to be furnished from other sources)	25,000
5. *Travel*, for purposes of liaison, conduct of field tests, visits to manufacturers and other institutions, etc.	6,000
6. *Miscellaneous*, communications, publications, reduction of data, etc.	1,000
Subtotal	$57,500
7. University Overhead (15% of subtotal)	8,625
Total Proposed	$66,125

D. Bibliography of Completed Work Pertaining to Proposal

[Citations of 31 relevant papers were given but are omitted in this reproduction.]

. . . . .

Submitted by J. A. Van Allen with copies to Kaplan, Joyce, Schilling and the IGY Technical Panel on Rocketry.

APPENDIX B

Correspondence Regarding *Pravda* Article, 1959

NEW YORK
Herald Tribune
A European Edition is Published Daily in Paris

PEnnsylvania 6-4000   230 West 41st Street, New York 36

March 12, 1959

Dr. James A. Van Allen
State University of Iowa
Iowa City, Iowa

Dear Dr. Van Allen:

    Enclosed is a translation of an article that appeared in the Moscow edition of Pravda as well as the piece itself which includes figures. These were sent to us by our Moscow correspondent.

    Some of the information in here seems contrary to what has been printed. In particular, the second page of the translation says the Soviets reported finding of the two zones of radiation to the IGY assembly last summer.

    I was under the impression that Pioneer III, fired last December, was the first vehicle to have discovered the radiation zones and that your laboratory, through the NASA, was first to announce this.

    I wonder if you could give me some comment on this matter and the article in general for a possible news story. I would appreciate hearing from you as soon as possible, and if the material in the article is news to you, I would be grateful if you did not mention it until I have a chance to do a story on it.

                                                 Very truly yours,

                                               Robert C. Toth
                                               Science Reporter

RCT:RL
Encls.

Department of Physics
State University of Iowa
Iowa City, Iowa

13 March 1959

Mr. Robert C. Toth
New York Herald Tribune
230 West 41st Street
New York 36, New York

Dear Mr. Toth:

1. Thank you for your letter of the 12th inst. and for the copy of the Pravda article of Friday, March 6, 1959 by S. N. Vernov and A. Chudakov. Dr. Ray at this laboratory has also translated this article and it has been discussed here with considerable interest during the past few days.

2. Let me first refer you to my recent account of our discovery of the trapped radiation by Explorer I and the subsequent development of our knowledge by Explorers III and IV and by Pioneers I and III as printed in "Scientific American" for March 1959, pp. 39–47. See also our recent article in the British journal "Nature", Vol. 183, pp. 430–434 of February 14, 1959. (The latter journal has rapid international distribution).

3. The first public announcement of the discovery of the existence of the trapped radiation was my paper before the joint session of the American Physical Society and the National Academy of Sciences on May 1, 1958. Up to that time there had been, to my knowledge, no announcement of Soviet radiation observations with Sputniks I and II, which gave any indication of the existence of the high intensity of the trapped radiation. Sputnik III was launched on May 15, 1958, two weeks later.

4. Dr. Ernest C. Ray, one of my collaborators in our satellite and space probe program, was one of the U.S. delegates to the Fifth General Assembly of C.S.A.G.I. (I.G.Y.) in Moscow, 30 July – 9 August 1958. He heard the papers given there by Vernov, Chudakov, and Lebedinsky and discussed their results and ours with them. The available results of Sputniks I and II were very fragmentary at that time as were those of Sputnik III. At no time did they suggest a picture of the overall situation resembling Figure 2 of the March 6 Pravda article.

5. In retrospect I should say that both the Soviets and we *independently* did have the basis—as of last August—for *speculating* on the structure of the radiation zone as shown in the references of paragraph 2 above. But neither group had the perspicacity to do so at that time.

6. We felt justified in reporting the existence of the two distinct zones only after the extensive and conclusive direct observations with Pioneer III. I reported our observations and conclusions at the December 27, 1958 public meeting of the American Astronautical Society in Washington, D. C. My diagram of the double structure of the radiation region was distributed to anyone who wished a copy at that time and the diagram was reproduced promptly in a number of newspapers and popular magazines (including the N.Y. Times).

7. The Soviet Cosmic Rocket "Mechta" was not launched until about a week later—namely on January 2, 1959. It was impossible for the Soviets to have had a sound observational basis for their Figure 2 of the March 6 Pravda article prior to the flight of Mechta,—unless, of course, they had previously made other, unannounced flights, of this nature.

8. Hence, I conclude that the statement of Vernov and Chudakov that their Figure 2 was reported in August 1958 to the I.G.Y. conference contains a considerable admixture of hindsight. Nonetheless, I wish to express my very considerable admiration for their work. And, as a scientist, I am delighted with the confirmation of our observations which the Soviets are obtaining.

9. Finally, you may be interested to know that Pioneer IV has provided a full confirmation of our Pioneer III results and has, in addition, yielded some very interesting new results which will be reported in the near future.

Sincerely yours,

J. A. Van Allen

Bibliography

An asterisk (*) indicates a reference work, not necessarily cited in the text, that the reader may wish to consult for a fuller explanation of magnetospheric physics.

Akasofu, S.-I., and S. Chapman
  1961. The Ring Current, Geomagnetic Distribution and the Van Allen Radiation Belts. *Journal of Geophysical Research*, 66:1321–50.
Alfvén, H.
  1950. *Cosmical Electrodynamics*. 237 pp. Oxford: Oxford University Press, Clarendon Press.
  1955. On the Electric Field Theory of Magnetic Storms and Aurorae. *Tellus*, 7:50–64.
Alfvén, H., and C.-G. Fälthammer
  1963. *Cosmical Electrodynamics*. 2nd Ed.. 228 pp. Oxford: Oxford University Press, Clarendon Press.
Allen, L., Jr., et al.
  1959. Project Jason Measurement of Trapped Electrons from a Nuclear Device by Sounding Rockets. *Journal of Geophysical Research*, 64:893–907. [Allen et al. = Allen, L., Jr., J. L. Beavers II, W. A. Whitaker, J. A. Welch, Jr., and R. O. Walton.]
Anderson, K. A.
  1958. Soft Radiation Events at High Altitude during the Magnetic Storm of August 29–30, 1957. *Physical Review*, 111:1397–1405.
Aono, Y., and K. Kawakami
  1958. Cosmic Rays Observed by Satellite 1958 Alpha. *Report of Ionosphere and Space Research in Japan*, 12:28–36.
Armstrong, A. H., et al.
  1961. Charged Particles in the Inner Van Allen Radiation Belt. *Journal of Geophysical Research*, 66:351–57. [Armstrong et al. = Armstrong, A. H., F. B. Harrison, H. H. Heckman, and L. Rosen.]
Arnoldy, R., R. Hoffman, and J. R. Winckler
  1960a. Solar Cosmic Rays and Soft Radiation Observed at 5,000,000 Kilometers from Earth. *Journal of Geophysical Research*, 65:3004–7.
  1960b. Measurements of the Van Allen Radiation Belts during Geomagnetic Storms. In *Space Research* (Proceedings of First International Space Science Symposium), ed. H. Kallmann Bijl,877–96. Amsterdam: North-Holland.
  1960c. Observations of the Van Allen Radiation Region during August and September 1959, Part 1. *Journal of Geophysical Research*, 65:1361–76.
  1962. Observations of the Van Allen Radiation Region during August and September 1959. Part 4. The Outer Zone Electrons. *Journal of Geophysical Research*, 67:2595–2612.
Arnoldy, R. L., et al.
  1962. Observations of the Van Allen Radiation Regions during August and September 1959. Part 5. Visual Auroras, High Altitude X-Ray Bursts and Simultaneous Satellite Observations. *Journal of Geophysical Research*, 67:3673–86. [Arnoldy et al. = Arnoldy, R. L., R. A. Hoffman, J. R. Winckler, and S.-I. Akasofu.]

Axford, W. I.,,
  1962. The Interaction between the Solar Wind and the Earth's Magnetosphere. *Journal of Geophysical Research*, 67:3791–96.
Axford, W. I., and C. O. Hines
  1961. A Unifying Theory of High-Latitude Geophysical Phenomena and Geomagnetic Storms. *Canadian Journal of Physics*, 39:1433–64.
Basler, R. P., R. N. Dewitt, and G. C. Reid
  1960. Radiation Information from 1958 δ 2 [Sputnik III]. *Journal of Geophysical Research*, 65:1135–38.
Beard, D. B.
  1960. The Interaction of the Terrestrial Magnetic Field with the Solar Corpuscular Radiation. *Journal of Geophysical Research*, 65:3559–68.
Berthold, W. K., A. K. Harris, and H. J. Hope
  1960. World-Wide Effects of Hydromagnetic Waves Due to Argus. *Journal of Geophysical Research*, 65:2233–39.
Biermann, L, and R. Lüst
  1966. The Interaction of the Solar Wind with Comets (Natural and Artificial). In *The Solar Wind*, ed. R. J. Mackin, Jr., and M. Neugebauer, 355–63, Pasadena, Calif.: Jet Propulsion Laboratory. [See also L. Biermann, *Zeitschrift für Astrophysik*, 29(1951):274.]
Birkeland, Kr.
  1908. *The Norwegian Aurora Polaris Expedition, 1902–1903*. Vol. 1, 1st sec., 1908,
  1913. 315 pp. and 21 plates, and 2d sec., 1913, 801 pp. and 42 plates. Kristiania: H. Aschehoug and Company.
Block, L.
  1955. Model Experiments on Aurorae and Magnetic Storms. *Tellus*, 7:65–86.
Bonetti, A., et al.
  1963. Explorer 10 Plasma Measurements. *Journal of Geophysical Research*, 68:4017–63. [Bonetti et al = Bonetti, A., H. S. Bridge, A. J. Lazarus, B. Rossi, and F. Scherb.]
Bowen, I. S., R. A. Millikan, and H. V. Neher
  1938. New Light on the Nature and Origin of the Incoming Cosmic Rays. *Physical Review*, 53:855–61.
Cahill, L. J., Jr.
  1959a. Investigation of the Equatorial Electrojet by Rocket Magnetometer. *Journal of Geophysical Research*, 64:489–503.
  1959b. Detection of an Electrical Current in the Ionosphere above Greenland. *Journal of Geophysical Research*, 64:1377–80.
Cahill, L. J., and P. G. Amazeen
  1963. The Boundary of the Geomagnetic Field. *Journal of Geophysical Research*, 68:1835–43.
Carpenter, D. L.
  1963. Whistler Evidence of a "Knee" in the Magnetospheric Ionization Density Profile. *Journal of Geophysical Research*, 68:1675–82.
Chamberlain, J. W.
  1961. *Physics of the Aurora and Airglow*. 704 pp. New York: Academic Press.
Chapman, S.
  1955. Physics and Chemistry of the Upper Atmosphere. 278 pp. Including Appendixes by others and Notes by E. C. Ray. Lecture notes, University of Iowa.
Chapman, S., and J. Bartels
  1940. *Geomagnetism*. vols. 1 and 2. 1049 pp. Oxford: Oxford University Press, Clarendon Press.
Chapman, S., and V. C. A. Ferraro
  1931. A New Theory of Magnetic Storms. *Terrestrial Magnetism and Atmospheric*
  1932. *Electricity*, 36:77, 171; 37:147, 421.

Chinburg, D. L.
 1960. Great Magnetic Storm of March 31–April 3, 1960. *Journal of Geophysical Research*, 65:2206–8.
Christofilos, N. C.
 1959. The Argus Experiment. *Journal of Geophysical Research*, 64:869–75.
Coleman, P. J., L. Davis, and C. P. Sonett
 1960. Steady Component of the Interplanetary Magnetic Field: Pioneer V. *Physical Review Letters*, 5:43–46.
Coleman, P. J., C. P. Sonett, and D. L. Judge
 1960. Some Preliminary Results of the Pioneer V Magnetometer Experiment. *Journal of Geophysical Research*, 65:1856–57.
Cox, A.
 1975. The Frequency of Geomagnetic Reversals and the Symmetry of the Nondipole Field. *Reviews of Geophysics and Space Physics*, 13:35–51.
Cullington, A. L.
 1958. A Man-Made or Artificial Aurora. *Nature (London)*, 182:1365–66.
Davis, L. R., O. E. Berg, and L. H. Meredith
 1960. Direct Measurements of Particle Fluxes in and near Auroras. In *Space Research* (Proceedings of First International Space Science Symposium), ed. H. Kallmann Bijl, 721–35. Amsterdam: North-Holland.
Davis, L. R., and D. B. Chang
 1962. On the Effects of Geomagnetic Fluctuations on Trapped Particles. *Journal of Geophysical Research*, 67:2169–79.
Davis, L. R., and J. M. Williamson
 1963. Low-Energy Trapped Protons. In *Space Research III* (Proceedings of Third International Space Science Symposium), ed. W. Priester, 365–75. Amsterdam: North-Holland.
Dessler, A. J.
 1967. *Solar Wind and Interplanetary Magnetic Field. *Reviews of Geophysics*, 5:1–41.
Dessler, A. J., and E. N. Parker
 1959. Hydromagnetic Theory of Geomagnetic Storms. *Journal of Geophysical Research*, 64:2239–52.
Dessler, A. J., and E. H. Vestine
 1960. Maximum Total Energy of Van Allen Radiation Belt. *Journal of Geophysical Research*, 65:1069–71.
Dolginov, S. Sh., L. N. Zhuzgov, and N.V. Pushkov.
 1960. Preliminary Report on Geomagnetic Measurements Carried Out from the Third Soviet Artificial Earth Satellite. In *Artificial Earth Satellites*, ed. L. V. Kurnosova, 2:63–67. New York: Plenum Press.
Dolginov, S. Sh., et. al.
 1960. Measuring the Magnetic Fields of the Earth and Moon by Means of Sputnik III and Space Rockets I and II. In *Space Research* (Proceedings of First International Space Science Symposium), ed. H. Kallmann Bijl, 863–68. Amsterdam: North-Holland. [Dolginov et al. = Dolginov, S. Sh., E. G. Eroshenko, L. N. Zhuzgov, N. V. Pushkov, and L. O. Tyurmina.]
 1961. Magnetic Measurements with the Second Cosmic Rocket. In *Artificial Earth Satellites*, ed. L.V. Kurnosova, 3/4/5:490–502. New York: Plenum Press. [Dolginov et al. = Dolginov, S. Sh., E. G. Eroshenko, L. N. Zhuzgov, N. V. Pushkov, and L.O. Tyurmina.] [See also Dolginov, S. Sh., and N. V. Pushkov, *Doklady Akademii Nauk, SSSR*, 129 (1959):77.]
Dungey, J. W.
 1958. *Cosmic Electrodynamics*. 183 pp. Cambridge: Cambridge University Press.
 1961. Interplanetary Magnetic Field and the Auroral Zone. *Physical Review Letters*, 6:47–48.

Ellis, R. A., Jr., M. B. Gottlieb, and J. A. Van Allen
  1954. Low Momentum End of the Spectra of Heavy Primary Cosmic Rays. *Physical Review*, 95:147–59.
Fan, C. Y., P. Meyer, and J. A. Simpson
  1960a. Preliminary Results from the Space Probe Pioneer V. *Journal of Geophysical Research*, 65:1862–63.
  1960b. Rapid Reduction of Cosmic-Radiation Intensity Measured in Interplanetary Space. *Physical Review Letters*, 5:269–71.
  1960c. Experiments on the Eleven Year Changes of Cosmic Ray Intensity Using a Space Probe. *Physical Review Letters*, 5:272–74.
  1960d. Trapped and Cosmic Radiation Measurements from Explorer VI. In *Space Research* (Proceedings of First International Space Science Symposium), ed. H. Kallmann Bijl, 951–66. Amsterdam: North-Holland.
  1961. Dynamics and Structure of the Outer Radiation Belt. *Journal of Geophysical Research*, 66:2607–40.
Forbush, S. E., G. Pizzella, and D. Venkatesan
  1962. The Morphology and Temporal Variations of the Van Allen Radiation Belt. *Journal of Geophysical Research*, 67:3651–68.
Forbush, S. E., D. Venkatesan, and C. E. McIlwain
  1961. Intensity Variations in the Outer Van Allen Radiation Belt. *Journal of Geophysical Research*, 66:2275–87.
Fowler, P. H., and C. J. Waddington
  1958. An Artificial Aurora. *Nature (London)*, 182:1728.
Frank, L. A.
  1962. Efficiency of a Geiger-Mueller Tube for Non-Penetrating Electrons. *Journal of the Franklin Institute*, 273:91–106.
  1967. On the Extraterrestrial Ring Current during Geomagnetic Storms. *Journal of Geophysical Research*, 72:3753–67.
Frank, L. A., and J. A. Van Allen
  1964. A Survey of Magnetospheric Boundary Phenomena. In *Research in Geophysics*, vol. 1, *Sun, Upper Atmosphere, and Space*, ed. H. Odishaw, 161–87. Cambridge, Mass.: MIT Press.
Frank, L. A., J. A. Van Allen, and J. D. Craven
  1964. Large Diurnal Variations of Geomagnetically Trapped and of Precipitated Electrons Observed at Low Altitudes. *Journal of Geophysical Research*, 69:3155–67.
Frank, L. A., J. A. Van Allen, and H. K. Hills
  1964. A Study of Charged Particles in the Earth's Outer Radiation Zone with Explorer 14. *Journal of Geophysical Research*, 69:2171–91.
Frank, L. A., J. A. Van Allen, and E. Macagno
  1963. Charged Particle Observations in the Earth's Outer Magnetosphere. *Journal of Geophysical Research*, 68:3543–54.
Frank, L. A., et al.
  1963. Absolute Intensities of Geomagnetically Trapped Particles. *Journal of Geophysical Research*, 68:1573–79. [Frank et al. = Frank, L. A., J. A. Van Allen, W. A. Whelpley, and J. D. Craven.]
Freden, S. C., and R. S. White
  1959. Protons in the Earth's Magnetic Field. *Physical Review Letters*, 3:9–10, 145.
  1960. Particle Fluxes in the Inner Radiation Belt. *Journal of Geophysical Research*, 65:1377–83.
Freeman, J. W., Jr.
  1964. The Morphology of the Electron Distribution in the Outer Radiation Zone and Near the Magnetospheric Boundary as Observed by Explorer 12. *Journal of Geophysical Research*, 69:1691–1723.

Freeman, J. W., J. A. Van Allen, and L. J. Cahill
- 1963. Explorer 12 Observations of the Magnetospheric Boundary and the Associated Solar Plasma on September 13, 1961. *Journal of Geophysical Research*, 68:2121–30.

Gangnes, A. V., J. F. Jenkins, Jr., and J. A. Van Allen
- 1949. The Cosmic-Ray Intensity above the Atmosphere. *Physical Review*, 75:57–69. Erratum 75:892.

Gilbert, W.
- 1600. *De Magnete, Magneticisque Corporibus, et de Magno Magnete Tellure; Physiologia Nova*. London. [English Translation by P. F. Mottelay, 368 pp. New York: John Wiley and Sons, 1893. Also New York: Dover Publications reprint, 1958.]

Gold, T.
- 1959a. Motions in the Magnetosphere of the Earth. *Journal of Geophysical Research*, 64: 1219–24.
- 1959b. Origin of the Radiation near the Earth Discovered by Means of Satellites. *Nature (London)*, 183:355–58.

Green, C. M., and M. Lomask
- 1970. *Vanguard—A History*. 308 pp. Washington, D.C.: National Aeronautics and Space Administration.

Gringauz, K. I., V. V. Bezrukikh, and V. D. Ozerov
- 1961. Results of Measurements of the Concentration of Positive Ions in the Atmosphere, Using Ion Traps Mounted on the Third Soviet Earth Satellite. In *Artificial Earth Satellites*, ed. L. V. Kurnosova, 6:77–121. New York: Plenum Press.

Gringauz, K. I., and S. M. Rytov
- 1960. Relationship between the Results of Measurements by Charged Particle Traps on the Soviet Cosmic Rockets and Magnetic Field Measurements on the American Satellite Explorer 6 and Rocket Pioneer 5. *Soviet Physics-Doklady*, 135:1225–28. [Originally *Doklady Akademii Nauk SSSR*, 135 (1960):48–51.]

Gringauz, K. I., et al.
- 1961a. Ionized Gas and Fast Electrons in the Earth's Neighborhood and Interplanetary Space. In *Artificial Earth Satellites*, ed. L. V. Kurnosova, 6:130–36. New York: Plenum Press. [Gringauz et al. = Gringauz, K. I., V. G. Kurt, V. G. Moroz, and I. S. Shklovoskii.]
- 1961b. A Study of the Interplanetary Ionized Gas, High Energy Electrons, and Corpuscular Radiation from the Sun by Means of the Three-Electrode Trap for Charged Particles on the Second Soviet Cosmic Rocket. *Soviet Physics-Doklady*, 5:361–64. [Originally *Doklady Akademii Nauk SSSR*, 131 (1960):1301–4.] [Gringauz et al. = Gringauz, K. I., V. V. Bezrukikh, V. D. Ozerov, and R. E. Rybchinskii.]
- 1963. On Results of Experiments with Charged Particle Traps in the Second Radiation Belt and in the Outermost Belt of Charged Particles. In *Space Research III* (Proceedings of Third International Space Science Symposium), ed. W. Priester, 432–37. Amsterdam: North-Holland. [Gringauz et al. = Gringauz, K. I., S. M. Balandina, G. A. Bordovsky, and N. M. Shutte.]
- 1964. Measurements Made in the Earth's Magnetosphere by Means of Charged Particle Traps aboard the Mars 1 Probe. In *Space Research IV* (Proceedings of the Fourth International Space Symposim), ed. P. Muller, 621–26. Amsterdam: North-Holland. [Gringauz et al. = Gringauz, K. I., V. V. Bezrukikh, L. S. Musatov, R. E. Rybchinsky, and S. M. Sheronova.]

Gurnett, D. A., and B. J. O'Brien
- 1964. High-Latitude Geophysical Studies with Satellite Injun 3. 5. Very Low-Frequency Electromagnetic Radiation. *Journal of Geophysical Research*, 69:65–89.

Heppner, J. P., et al.
- 1963. Explorer 10 Magnetic Field Measurements. *Journal of Geophysical Research*, 68:1–46. [Heppner et al. = Heppner, J. P., N. F. Ness, C. S. Scearce, and T. L. Skillman.]

Herlofson, N.
- 1960. Diffusion of Particles in the Earth's Radiation Belts. *Physical Review Letters*, 5:414–16.

Herz, A. J., et al.
- 1960. Radiation Observations with Satellite 1958 δ [Sputnik III] over Australia. In *Proceedings of the Moscow Cosmic Ray Conference*, ed. S. I. Syrovatsky, 3:32–40. Moscow: International Union of Pure and Applied Physics. [Herz et al. = Herz, A. J., K. W. Ogilvie, J. Olley, and R. B. White.]

Hess, W. N.
- 1962. Energetic Particles in the Inner Van Allen Belt. *Space Science Reviews*, 1:278–312.
- 1968. *The Radiation Belt and Magnetosphere*, 548 pp. Waltham, Mass.: Blaisdell Publishing Company.

Hoffman, R. A.
- 1961. Observations of the Van Allen Radiation Regions during August and September 1959. Part 2. The Capetown Anomaly and the Shape of the Outer Belt. *Journal of Geophysical Research*, 66:4003–6.

Hoffman, R. A., R. L. Arnoldy, and J. R. Winckler
- 1962a. Observations of the Van Allen Radiation Regions during August and September 1959. Part 3. The Inner Belt. *Journal of Geophysical Research*, 67:1–12.
- 1962b. Observations of the Van Allen Radiation Regions during August and September 1959. Part 6. Properties of the Outer Region. *Journal of Geophysical Research*, 67:4543–76.

Holliday, C. T.
- 1950. Seeing the Earth from 80 Miles Up. *National Geographic Society*, 98 (4):511–28.

Holly, F. E., and R. A. Johnson
- 1960. Measurement of Radiation in the Lower Van Allen Belt. *Journal of Geophysical Research*, 65:771–72.

Hopfield, J. J., and H. E. Clearman, Jr.
- 1948. The Ultraviolet Spectrum of the Sun from V-2 Rockets. *Physical Review*, 73:877–84.

Hovestadt, D., et al.
- 1978. Evidence for Solar Wind Origin of the Energetic Heavy Ions in the Earth's Radiation Belt. *Geophysical Research Letters*, 5:1055–57. [Hovestadt et al. = Hovestadt, D., G. Gloeckler, C. Y. Fan, L. A. Fisk, F. M. Ipavich, B. Klecker, J. J. O'Gallagher, and M. Scholer.]

Jánossy, L.
- 1948. *Cosmic Rays*. 424 pp. Oxford: Oxford University Press, Clarendon Press.

Johnson, F. S.
- 1960. The Gross Character of the Geomagnetic Field in the Solar Wind. *Journal of Geophysical Research*, 65:3049–51.

Josias, C.
- 1959. Pioneer's Radiation-Detecting Instrument. *Astronautics*, July, pp. 32–33, 114–15.

Kellogg, P. J.
- 1959a. Possible Explanation of the Radiation Observed by Van Allen at High Altitudes in Satellites. *Il Nuovo Cimento*, ser. X, 11:48–66.
- 1959b. Van Allen Radiation of Solar Origin. *Nature (London)*, 183:1295–97.
- 1960. Electrons of the Van Allen Radiation. *Journal of Geophysical Research*, 65:2705–13.
- 1962. Flow of Plasma around the Earth. *Journal of Geophysical Research*, 67:3805–11.

Kellogg, P. J., E. P. Ney, and J. R. Winckler
  1959. Geophysical Effects Associated with High Altitude Explosions. *Nature (London)*, 183:358–61.
Kennel, C. F., and H. E. Petschek
  1966. A Limit on Stably Trapped Particle Fluxes. *Journal of Geophysical Research*, 71:1–28.
Klass, P. J.
  1955. U.S. Plans to Launch 12 Earth Satellites. *Aviation Week*, 63(25):12–13.
Krassovskii, V. I.
  1960. Results of Scientific Investigations Made by Soviet Sputniks and Cosmic Rockets. *Astronautica Acta*, 6:32–47.
Krassovskii, V. I., et al.
  1960. On Fast Corpuscles in the Upper Atmosphere. In *Proceedings of the Moscow Cosmic Ray Conference*, ed. S. I. Syrovatsky, 3:59–63. Moscow: International Union of Pure and Applied Physics. [Krassovskii et al. = Krassovskii, V. I., I. S. Shklovskii, G. I. Galperin, and E. M. Svetlitsky.]
  1961. Discovery of Approximately 10-kev Electrons in the Upper Atmosphere. In *Artificial Earth Satellites*, ed. L. V. Kurnosova, 6:137–55. New York: Plenum Press. [Krassovskii et al. = Krassovskii, V. I., I. S. Shklovskii, Yu. I. Galperin, E. M. Svetlitskii, Yu. M. Kushnir, and G. A. Bordovskii.]
Krieger, F. J.
  1958. *Behind the Sputniks*. 380 pp. Washington, D.C.: Public Affairs Press.
Kurnosova, L. V., et al.
  1961. Cosmic Radiation Studies during the Flight of the Second Lunar Rocket. In *Artificial Earth Satellites*, ed. L. V. Kurnosova, 3/4/5:512–23. New York: Plenum Press. [Kurnosova et al. = Kurnosova, L. V., V. I. Logachev, L. A. Razorenov, and M. I. Fradkin.]
Leiphart, J. P.
  1962. Penetration of the Ionosphere by Very-Low-Frequency Radio Signals—Interim Results of the Lofti 1 Experiment. *Proceedings of the Institute of Radio Engineers*, 50:6.

Lin, W. C., and J. A. Van Allen
  1964. Observation of Solar Cosmic Rays from October 13, 1959 to February 17, 1961 with Explorer VII. In *Space Exploration and the Solar System*, ed. B. Rossi, 194–235. New York: Academic Press.
Lin, W. C., D. Venkatesan, and J. A. Van Allen
  1963. Latitude Survey of Cosmic-Ray Intensity by Explorer 7, October 1959 to February 1961. *Journal of Geophysical Research*, 68:4885–96.
Loftus, T. A.
  1969. Disturbance of the Inner Van Allen Belt as Observed by Explorer I. 49 pp. M.S. thesis, University of Iowa.
Lowrie, J. A., V. B. Gerard, and P. J. Gill
  1959. Magnetic Effects Resulting from Two High-Altitude Nuclear Explosions. *Nature (London)*, 184:34, 51–52.
Ludwig, G. H.
  1959. Cosmic-Ray Instrumentation in the First U.S. Earth Satellite. *Review of Scientific Instruments*, 30:223–29.
  1961. The Instrumentation in Earth Satellite 1958 Gamma [Explorer III]. *IGY Satellite Report*, 13:31–93. Washington, D. C.: National Academy of Sciences. [Also State University of Iowa Research Report 59-3, February 1959.]
Ludwig, G. H., and J. A. Van Allen
  1956. Instrumentation for a Cosmic Ray Experiment for the Minimal Earth Satellite. State University of Iowa memo, May 30.

Ludwig, G. H., and W. A. Whelpley
: 1960. Corpuscular Radiation Experiment of Satellite 1959 Iota (Explorer VII). *Journal of Geophysical Research*, 65:1119–24.

Maeda, H.
: 1959. Geomagnetic Disturbance Due to Nuclear Explosion. *Journal of Geophysical Research*, 64:863–64.

Malmfors, K. G.
: 1945. Determination of Orbits in the Field of a Magnetic Dipole with Applications to the Theory of the Diurnal Variation of Cosmic Radiation. *Arkiv Matematik, Astronomie och Fysik*, 32A(8):1–64.

Malville, J. M.
: 1959. Artificial Auroras Resulting from the 1958 Johnston Island Nuclear Explosions. *Journal of Geophysical Research*, 64:2267–70.

Mason, R. G., and M. J. Vitousek
: 1959. Some Geomagnetic Phenomena Associated with Nuclear Explosions. *Nature (London)*, 184:52–54.

Matsushita, S.
: 1959. On Artificial Geomagnetic and Ionospheric Storms Associated with High-Altitude Detonations. *Journal of Geophysical Research*, 64:1149–61.

Maxwell, J. C.
: 1891. *A Treatise on Electricity and Magnetism*. 3rd ed. vol. 1, 506 pp. and 13 plates; vol. 2, 500 pp. and 7 plates. [Reprinted by New York: Dover Publications, 1954.]

McIlwain, C. E.
: 1956. Cosmic Ray Intensity above the Atmosphere at Northern Latitudes. 50 pp. M.S. thesis, University of Iowa.
: 1960a. Scintillation Counters in Rockets and Satellites. *Institute of Radio Engineers Transactions in Nuclear Science*, NS7:159–64.
: 1960b. Direct Measurement of Particles Producing Visible Auroras. *Journal of Geophysical Research*, 65:2727–47.
: 1961. Coordinates for Mapping the Distribution of Magnetically Trapped Particles. *Journal of Geophysical Research*, 66:3681–91.

McNish, A. G.
: 1959. Geomagnetic Effects of High-Altitude Nuclear Explosions. *Journal of Geophysical Research*, 64:2253–65.

Meredith, L. H., M. B. Gottlieb, and J. A. Van Allen
: 1955. Direct Detection of Soft Radiation above 50 Kilometers in the Auroral Zone. *Physical Review*, 97:201–5.

Meredith, L. H., J. A. Van Allen, and M. B. Gottlieb
: 1955. Cosmic-Ray Intensity above the Atmosphere at High Latitudes. *Physical Review*, 99:198–209

Mitra, S. K.
: 1948. *The Upper Atmosphere*. 616 pp. Calcutta: The Royal Asiatic Society of Bengal.

Miyazaki, Y., and H. Takeuchi
: 1958. Altitude Dependence and Time Variation of the Radiation Observed by U. S. Satellite 1958α [Explorer I]. *Report of Ionosphere and Space Research in Japan*, 12:448–58.
: 1960. Radiation Measurements from Satellite 1958 Epsilon [Explorer IV]. In *Space Research* (Proceedings of First International Space Science Symposium), ed. H. Kallmann Bijl, 869–76. Amsterdam: North-Holland.

Montgomery, D. J. X.
: 1949. *Cosmic Ray Physics*. 370 pp. Princeton, N.J.: Princeton University Press.

Morrison, P.
: 1956. Solar Origin of Cosmic-Ray Time Variations. *Physical Review*, 101:1397–1404.

Naugle, J. E., and D. A. Kniffen
- 1961 The Flux and Energy Spectrum of the Protons in the Inner Van Allen Belt. *Physical Review Letters*, 7:3–6.
- 1963. Variations of the Proton Energy Spectrum with Position in the Inner Van Allen Belt. *Journal of Geophysical Research*, 68:4065–89.

Newell, H. E., Jr.
- 1953. *High Altitude Rocket Research*. 298 pp. New York: Academic Press.
- 1959. *Editor. Sounding Rockets*. 334 pp. New York: McGraw-Hill.
- 1980. *Beyond the Atmosphere*. 497 pp. Washington, D.C.: National Aeronautics and Space Administration.

Newman, P.
- 1959. Optical, Electromagnetic, and Satellite Observations of High-Altitude Nuclear Detonations. Part I. *Journal of Geophysical Research*, 64:923–32.

Newton, I.
- 1687. *Philosophiae Naturalis Principia Mathematica*. London: Royal Society Press, with imprimatur of S. Pepys.

Northrop, T. G.
- 1963. *The Adiabatic Motion of Charged Particles*. 109 pp. New York: Interscience Publishers.

Northrop, T. G., and E. Teller
- 1960. Stability of the Adiabatic Motion of Charged Particles in the Earth's Field. *Physical Review*, 117:215–25.

O'Brien, B. J.
- 1962. Lifetimes of Outer-Zone Electrons and Their Precipitation into the Atmosphere. *Journal of Geophysical Research*, 67:3687–3706.

O'Brien, B. J., and C. D. Laughlin
- 1962. An Extremely Intense Electron Flux at 1000 km Altitude in the Auroral Zone. *Journal of Geophysical Research*, 67:2667–72.

O'Brien, B. J., et al.
- 1962a. Absolute Electron Intensities in the Heart of the Earth's Outer Radiation Zone. *Journal of Geophysical Research*, 67:397–403. [O'Brien et al. = O'Brien, B. J., J. A. Van Allen, C. D. Laughlin, and L. A. Frank.]
- 1962b. Measurements of the Intensity and Spectrum of Electrons at 1000 km Altitude and High Latitudes. *Journal of Geophysical Research*, 67:1209–25. [O'Brien et al. = O'Brien, B. J., C. D. Laughlin, J. A. Van Allen, and L. A. Frank.]

Parker, E. N.
- 1958. Interaction of the Solar Wind with the Geomagnetic Field. *Physics of Fluids*, 1:171–87.
- 1960. Geomagnetic Fluctuations and the Form of the Outer Zone of the Van Allen Radiation Belt. *Journal of Geophysical Research*, 65:3117–30.
- 1963. *Interplanetary Dynamical Processes*. 272 pp. New York: Interscience Publishers.

Peterson, A. M.
- 1959. Optical, Electromagnetic, and Satellite Observations of High Altitude Nuclear Detonations. Part II. *Journal of Geophysical Research*, 64:933–38.

Piddington, J. H.
- 1960. Geomagnetic Storm Theory. *Journal of Geophysical Research*, 65:93–106.
- 1969. *Cosmic Electrodynamics*. 305 pp. New York: Wiley-Interscience.

Pizzella, G., C. E. McIlwain, and J. A. Van Allen
- 1962. Time Variations of Intensity in the Earth's Inner Radiation Zone, October 1959 through December 1960. *Journal of Geophysical Research*, 67:1235–53.

Porter, R. W.
- 1959. Symposium on Scientific Effects of Artificially Introduced Radiations at High Altitudes: Introductory Remarks. *Journal of Geophysical Research*, 64:865–67.

Richter, H. L., Jr., et al.
  1959. Instrumenting the Explorer I Satellite. *Electronics*, 32:39–43. [Richter et al. = Richter, H. L, Jr., W. Pilkington, J. P. Eyrand, W. S. Shipley, and L. W. Randolph.]
Rikitake, T.
  1966. *Electromagnetism and the Earth's Interior*. 308 pp. Amsterdam: Elsevier Publishing Co.
Roederer, J. G.
  1970. *Dynamics of Geomagnetically Trapped Radiation*. 166 pp. Berlin: Springer-Verlag.
Rosen, A., P. J. Coleman, Jr., and C. P. Sonett
  1959. Ionizing Radiation Detected by Pioneer II. *Planetary and Space Science*, 1:343–46.
Rosen, A., and T. A. Farley
  1961. Characteristics of the Van Allen Radiation Zones as Measured by the Scintillation Counter on Explorer VI. *Journal of Geophysical Research*, 66:2013–28.
Rosen, A., T. A. Farley, and C. P. Sonett
  1960. Soft Radiation Measurements on Explorer VI Earth Satellite. In *Space Research* (Proceedings of First International Space Science Symposium), ed. H. Kallmann Bijl, 938–50. Amsterdam: North-Holland.
Rosen, A., et al.
  1959. Ionizing Radiation at Altitudes of 3,500 to 3,600 Kilometers, Pioneer I. *Journal of Geophysical Research*, 64:709–12. [Rosen et al. = Rosen, A., C. P. Sonett, P. J. Coleman, Jr., and C. E. McIlwain.]
Rosenbluth, M. N., and C. L. Longmire
  1957. Stability of Plasmas Confined by Magnetic Fields. *Annals of Physics*, 1:120–40.
Rosser, W. G. V., et al.
  1962. Electrons in the Earth's Outer Radiation Zone. *Journal of Geophysical Research*, 67:4533–42. [Rosser et al. = Rosser, W. G. V., B. J. O'Brien, J. A. Van Allen, L. A. Frank, and C. D. Laughlin.]
Rossi, B.
  1963. Interplanetary Plasma. In *Space Research III* (Proceedings of Third International Space Science Symposium), ed. W. Priester, 529–39. Amsterdam: North-Holland.
Rossi, B., and S. Olbert
  1970. *Introduction to the Physics of Space*, 454 pp. New York: McGraw-Hill Book Co.
Rothwell, P., and C. E. McIlwain
  1960. Magnetic Storms and the Van Allen Radiation Belts: Observations with Satellite 1958ε (Explorer IV). *Journal of Geophysical Research*, 65:799–806.
Schulz, M., and L. J. Lanzerotti
  1974. *Particle Diffusion in the Radiation Belts*. 215 pp. New York: Springer-Verlag.
Singer, S. F.
  1957. A New Model of Magnetic Storms and Aurorae. *Transactions of the American Geophysical Union*, 38:175–90.
  1958a. "Radiation Belt" and Trapped Cosmic Ray Albedo. *Physical Review Letters*, 1:171–73.
  1958b. Trapped Albedo Theory of the Radiation Belt. *Physical Review Letters*, 1:181–83.
Singer, S. F., E. Maple, and W. A. Bowen
  1951. Evidence for Ionospheric Currents from Rocket Experiments near the Geomagnetic Equator. *Journal of Geophysical Research*, 56:265–81.
Snyder, C. W.
  1959. The Upper Boundary of the Van Allen Radiation Belts. *Nature (London)*, 184:439–40.
Snyder, C. W., M. Neugebauer, and U. R. Rao
  1963. The Solar Wind Velocity and Its Correlation with Cosmic-Ray Variations and

with Solar and Geomagnetic Activity. *Journal of Geophysical Research*, 68:6361–70.

Sonett, C. P., E. J. Smith, and A. R. Sims
  1960. Surveys of the Distant Geomagnetic Field: Pioneer I and Explorer VI. In *Space Research* (Proceedings of First International Space Science Symposium), ed. H. Kallmann Bijl, 921–37. Amsterdam: North-Holland.

Sonett, C. P., et al.
  1960a. A Radial Rocket Survey of the Distant Geomagnetic Field. *Journal of Geophysical Research*, 65:55–58. [Sonett et al. = Sonett, C. P., D. L. Judge, A. R. Sims, and J. M. Kelso.]
  1960b. Current Systems in the Vestigal Geomagnetic Field: Explorer VI. *Physical Review Letters*, 4:161–63. [Sonett et al. = Sonett, C. P., E. J. Smith, D. L. Judge, and P. J. Coleman, Jr.]

Spitzer, L., Jr.
  1952. Equations of Motion for an Ideal Plasma. *Astrophysical Journal*, 116:299–316.

Spreiter, J. R., and B. R. Briggs
  1962. Theoretical Determination of the Form of the Boundary of the Solar Corpuscular Stream Produced by Interaction with the Magnetic Dipole Field of the Earth. *Journal of Geophysical Research*, 67:37–51.

Steiger, W. H., and S. Matsushita
  1960. Photographs of the High-Altitude Nuclear Explosion "Teak." *Journal of Geophysical Research*, 65:545–50.

Størmer, C.
  1907. Sur des trajectoires des corpuscles electrises dans l'space sons l'action du magnetisme terrestre, chapitre 4, *Archives des sciences physiques et naturelles*, 24:317–64.
  1955. *The Polar Aurora*. 403 pp. Oxford: Oxford University Press, Clarendon Press.

Townsend, J. W., Jr., et al.
  1959. The Aerobee-Hi Rocket. In *Sounding Rockets*, ed. H. E. Newell, Jr., 71–95. New York: McGraw-Hill. [Townsend et al. = Townsend, J. W., Jr., E. Pressly, R. M. Slavin, and L. Kraff, Jr.]

Treiman, S. B.
  1953. The Cosmic-Ray Albedo. *Physical Review*, 91:957–59.

Troitskaya, V. A.
  1960. Effects in Earth Currents Caused by High-Altitude Atomic Explosions. *Izvestiya Akademii Nauk, SSSR*, Seriya Geofizicheskaya, no. 9, 1321–29.

Van Allen, J. A.
  1953. The Cosmic Ray Intensity near the Geomagnetic Pole. *Il Nuovo Cimento*, 10:630–47.
  1956. *Editor. Scientific Uses of Earth Satellites*. 316 pp. Ann Arbor: University of Michigan Press.
  1957. Direct Detection of Auroral Radiation with Rocket Equipment. *Proceedings of the National Academy of Sciences*, 43:57–62.
  1959a. Balloon-launched Rockets for High-Altitude Research. In *Sounding Rockets*, ed. H. E. Newell, Jr., 143–64. New York: McGraw-Hill.
  1959b. The Geomagnetically-Trapped Corpuscular Radiation. *Journal of Geophysical Research*, 64:1683–89.
  1959c. Radiation Belts around the Earth. *Scientific American*, 200:39–47.
  1960. On the Radiation Hazards of Space Flight. In *The Physics and Medicine of the Atmosphere and Space*, ed. O. O. Benson and H. Strughold, 1–13. New York: John Wiley and Sons.
  1961. Observations of High Intensity Radiation by Satellite 1958 Alpha and Gamma [Explorers I and III]. *IGY Satellite Report*, 13:1–22. Washington, D.C.: National Academy of Sciences. [Reprinted with minor editing in *Space Science Comes of*

Age, ed. P. A. Hanle and V. D. Chamberlain, 58–75. Washington, D.C.: Smithsonian Institution Press, 1981.]
- 1962. Dynamics, Composition and Origin of the Geomagnetically-Trapped Corpuscular Radiation (Invited Discourse, 11th General Assembly of the International Astronomical Union, August 16, 1961). *Transactions of the International Astronomical Union*, 11B:99–136.
- 1964. Lifetimes of Geomagnetically-Trapped Electrons at Several MeV Energy. *Nature (London)*, 203:1006–7.

Van Allen, J. A., and L. A. Frank
- 1959a. Radiation around the Earth to a Radial Distance of 107,400 Kilometers. *Nature (London)*, 183:430–34.
- 1959b. Radiation Measurements to 658,300 km with Pioneer IV. *Nature (London)*, 184:219–24.

Van Allen, J. A., L. W. Fraser, and J. F. R. Floyd
- 1948. The Aerobee Sounding Rocket—A New Vehicle for Research in the Upper Atmosphere. *Science*, 108:746–47.

Van Allen, J. A., and A. V. Gangnes
- 1950a. The Cosmic-Ray Intensity above the Atmosphere at the Geomagnetic Equator. *Physical Review*, 78:50–52.
- 1950b. On the Azimuthal Asymmetry of Cosmic-Ray Intensity above the Atmosphere at the Geomagnetic Equator. *Physical Review*, 79:51–53.

Van Allen, J. A., and M. B. Gottlieb
- 1954. The Inexpensive Attainment of High Altitudes with Balloon-Launched Rockets. In *Rocket Exploration of the Upper Hemisphere*, ed. R. L. F. Boyd, M. J. Seaton, and H. S. W. Massey, 53–64. New York: Interscience Publishers.

Van Allen, J. A., and W. C. Lin
- 1960. Outer Radiation Belt and Solar Proton Observations with Explorer VII during March-April 1960. *Journal of Geophysical Research*, 65:2998–3003.

Van Allen, J. A., C. E. McIlwain, and G. H. Ludwig
- 1959a. Satellite Observations of Electrons Artificially Injected into the Geomagnetic Field. *Journal of Geophysical Research*, 64:877–91.
- 1959b. Radiation Observations with Satellite 1959 Epsilon [Explorer IV]. *Journal of Geophysical Research*, 64:271–86.

Van Allen, J. A., B. A. Randall, and S. M. Krimigis
- 1970. Energetic Carbon, Nitrogen and Oxygen Nuclei in the Earth's Outer Radiation Zone. *Journal of Geophysical Research*, 75:6085–91.

Van Allen, J. A., and E. C. Ray
- 1958. Progress Report on Cosmic Ray Observations in Satellite 1958 Alpha [Explorer I]. State University of Iowa, February 28, 1958.

Van Allen, J. A., and S. F. Singer
- 1952. Apparent Absence of Low Energy Cosmic-Ray Primaries. *Nature (London)*, 170:62–63.

Van Allen, J. A., and H. E. Tatel
- 1948. The Cosmic Ray Counting Rate of a Single Geiger Counter from Ground Level to 161 Kilometers Altitude. *Physical Review*, 73:245–51.

Van Allen, J. A., J. W. Townsend, Jr., and E. C. Pressly
- 1959. The Aerobee Rocket. In *Sounding Rockets*, ed. H. E. Newell, Jr., 54–70. New York: McGraw-Hill.

Van Allen, J. A., et al.
- 1958. Observation of High Intensity Radiation by Satellites 1958 Alpha and Gamma [Explorers I and III]. *Jet Propulsion*, September, pp. 588–92. [Van Allen et al.-Van Allen, J. A., G. H. Ludwig, E. C. Ray, and C. E. McIlwain.]

Vernov, S. N., and A. E. Chudakov
- 1960. Terrestrial Corpuscular Radiation and Cosmic Rays. In *Space Research* (Pro-

ceedings of First International Space Science Symposium), ed. H. Kallmann Bijl, 751–96. Amsterdam: North-Holland.

Vernov, S. N., et al.
- 1958a. Research on Variations of Cosmic Radiation. In *Annals of the International Geophysical Year*, ed. L. V. Berkner, G. Reid, J. Hanessian, Jr., and L. Cormeir, vol. 6, *Manual of Rockets and Satellites*, 263–75. London: Pergamon Press [Translated from *Uspekhi Fizicheskikh Nauk Akademii Nauk Sviuza SSSR*, 63(1957):149–62.] [Vernov et al. = Vernov, S. N., Y. I. Logachev, A. Y. Chudakov, and Y. G. Shafer.]
- 1958b. Artificial Satellite Measurements of Cosmic Radiation. *Doklady Akademii Nauk SSSR*, 120:1231–33 [Reprinted in *Artificial Earth Satellites*, ed. L. V. Kurnosova, 1:5–9. New York, Plenum Press, 1960.] [Vernov et al. = Vernov, S. N., N. I. Grigorov, Y. I. Logachev, and A. Y. Chudakov.]
- 1959a. Study of the Cosmic-Ray Soft Component by the 3rd Soviet Earth Satellite. *Planetary and Space Science*, 1:86–93. [Vernov et al. = Vernov, S. N., A. E. Chudakov, E. V. Gorchakov, J. L. Logachev, and P. V. Vakulov.]
- 1959b. Study of Terrestrial Corpuscular Radiation and Cosmic Rays during Flight of the Cosmic Rocket. *Doklady Akademii Nauk SSSR*, 125:304–7. [Vernov et al. = Vernov, S. N., A. Ye. Chudakov, P. V. Vakulov, and Yu. I. Logachev.]
- 1961. Radiation Measurements during the Flight of the Second Lunar Rocket. In *Artificial Earth Satellites*, ed. L. V. Kurnosova, 3/4/5:503–11. New York: Plenum Press. [Translated from *Doklady Akademii Nauk SSSR*, 130(1960):517–20.] [Vernov et al. = Vernov, S. N., A. E. Chudakov, P. V. Valukov, Yu. I. Logachev, and A. G. Nikolaev.]

Welch, J. A., Jr., and W. A. Whitaker
- 1959. Theory of Geomagnetically Trapped Electrons from an Artificial Source. *Journal of Geophysical Research*, 64:909–22.

White, J. R.
- 1973. High-Energy Proton Radiation Belt. *Reviews of Geophysics and Space Physics*, 11:595–632.

Winckler, J. R., P. D. Bhavsar, and K. A. Anderson
- 1962. A Study of Precipitation of Energetic Electrons from the Geomagnetic Field during Magnetic Storms. *Journal of Geophysical Research*, 67:3717–36.

Winckler, J. R., and L. Peterson
- 1957. Large Auroral Effect on Cosmic-Ray Detectors Observed at 8 g/cm^2 Atmospheric Depth. *Physical Review*, 108:903–4.

Winckler, J. R., et al.
- 1958. X rays from Visible Aurorae at Minneapolis. *Physical Review*, 110:1221–31. [Winckler et al. = Winckler, J. R., L. Peterson, R. Arnoldy, and R. Hoffman.]

Yagoda, H.
- 1960. Star Production by Trapped Protons in the Inner Radiation Belt. *Physical Review Letters*, 5:17–18.

Yoshida, S., G. H. Ludwig, and J. A. Van Allen
- 1960. Distribution of Trapped Radiation in the Geomagnetic Field. *Journal of Geophysical Research*, 65:807–13.

About the Author

James A. Van Allen is one of the world's pioneers in scientific research at high altitudes and in space, having been in the forefront of the work since 1945. In 1951 he was appointed professor of physics and head of the Department of Physics (since 1959, the Department of Physics and Astronomy) at the University of Iowa, positions that he has held continuously since then.

In early 1958 Dr. Van Allen and his students at Iowa discovered the radiation belts of the earth using equipment that they had designed, built, and flown on *Explorer I*, the first American satellite to orbit the earth. He has been the principal investigator for numerous flights of scientific equipment carried on high-altitude balloons and rockets and on twenty-four space missions, including satellites of the earth and the moon and the first missions to the planets Venus, Mars, Jupiter, and Saturn.

Dr. Van Allen is the editor of *Scientific Uses of Earth Satellites* and the author or co-author of more than two hundred papers and chapters of books on nuclear physics, auroral physics, cosmic rays, solar x rays, and the magnetospheres of the earth, Jupiter, and Saturn. The preparation of the first draft of *Origins of Magnetospheric Physics* was made possible by a Regents' Fellowship of the Smithsonian Institution, which Dr. Van Allen held at the National Air and Space Museum from January to August 1981.